Instructor's Guide

Solutions Manual/
Test Item File

to accompany

BIOCEMISTRY

Fourth Edition

Geoffrey Zubay
Columbia University

Instructor's Guide Prepared by
Geoffrey L. Zubay

Solutions Manual Prepared by
Larry Loomis-Price
Gwen Shafer

Test Item File Prepared by
Hugh Akers
Cindy Klevickis

Boston   Burr Ridge, IL   Dubuque, IA   Madison, WI   New York   San Francisco   St. Louis
Bangkok   Bogotá   Caracas   Lisbon   London   Madrid
Mexico City   Milan   New Delhi   Seoul   Singapore   Sydney   Taipei   Toronto

WCB/McGraw-Hill

A Division of The McGraw·Hill Companies

Instructor's Guide/Solutions Manual/Test Item File to accompany
BIOCHEMISTRY

Copyright © 1998 by The McGraw-Hill Companies, Inc. All rights reserved.
Previous editions 1983, 1988, 1993 by Wm. C. Brown Communications, Inc.
Printed in the United States of America.

This book is printed on acid-free paper.

1 2 3 4 5 6 7 8 9 0 QPD/QPD 9 0 9 8

ISBN 0-697-21902-X

www.mhhe.com

INSTRUCTOR'S GUIDE

Prepared by

Geoffrey L. Zubay

Boston Burr Ridge, IL Dubuque, IA Madison, WI New York San Francisco St. Louis
Bangkok Bogotá Caracas Lisbon London Madrid
Mexico City Milan New Delhi Seoul Singapore Sydney Taipei Toronto

INTRODUCTION

Biochemistry is a growing discipline, closely linked to several other fields of study. As biochemistry has grown, so has the need to relate biochemical phenomena to cell biology, physiology, and genetics. We must familiarize the student with the relationship of biochemistry to these other disciplines without going too far afield. The goal of this edition of *Biochemistry* is to provide a comprehensive, up-to-date teaching text that will enlighten today's students and equip them to deal with tomorrow's problems in medicine and other related areas of endeavour. As a discipline in college courses, graduate schools, and medical schools, biochemistry is of ever-increasing importance. The demands for a uniformly authoritative text have never been greater.

A unique aspect of our text in a field that is crowded with many texts is the team-of-experts approach. This text is written by experts whose knowledge for their chapters comes from the primary scientific literature and their scientific experiences. All of the contributors to our text are researchers who have made their mark in specific areas of biochemistry; they are also teachers of the subject. Their contributed chapters have been scrutinized for clarity of exposition and to insure that they fit into a coordinated final product.

Organizational Changes in the Fourth Edition

Our efforts in this edition have focused on integrating new knowledge into an already extensive body of information. Changes have been made throughout the text, with major changes being confined to areas in which the most important advances have been made. In some cases, old references have been removed from the ends of chapters, and in most cases, appropriate new references have been included.

The contents of the different parts and chapters should be self-evident from the titles and section headings. Parts 1 and 3 and the chapters therein are organized in the same way as they were in the third edition. In the remaining parts of this edition, substantial changes in organization have been made. In the third edition, part 2 covered the structures and functions of most of the major components of the cell. In the fourth edition, part 2 focuses on protein structure and function, leaving the discussion of structure and function of the remaining components to be dealt with immediately before addressing their metabolism. In this way the structure and function will be fresh in the mind as one delves into the metabolism, a clear advantage for the student.

In parts 4 and 5 of the third edition we attempted to divide the metabolism into a discussion of the catabolism and the anabolism, respectively. These parts have been supplanted by parts 4, 5, and 6 in the fourth edition, in which the intermediary metabolism is considered first for carbohydrates (part 4), then for lipids (part 5), and finally for nitrogen-containing compounds (part 6). This arrangement is considered superior because it facilitates comparison of the similar and dissimilar features of the metabolism for opposing anabolic and catabolic pathways and a discussion of their regulation, which usually focuses on strategies that prevent opposing pathways from functioning simultaneously.

The strategy of presenting structures and functions adjacent to the related metabolism sections is carried over to part 7, in which the structures and functions of nucleic acids are considered just prior to a discussion of their metabolism.

Although we believe that many instructors of biochemistry will find these changes in organization to be superior for teaching purposes, we also recognize that there are others who prefer the organizational scheme followed in the third edition. In consideration of this, the relevant materials have been flexibly packaged so that the order of teaching need not follow the order in the text. For example, if the instructor wishes to consider most of the structures first before getting into the metabolism, then after a consideration of part 2 as it appears in the fourth edition, one can turn to chapter 13 for a discussion of the structures of sugars and energy storage polysaccharides, followed by chapter 19 for a discussion of the structure and function of biological membranes, perhaps chapter 20, which deals with membrane transport, and, finally, chapter 30 for a discussion of the structures of DNA and nucleoproteins.

Physiological biochemistry, which comprised part 7 in the third edition, has been eliminated in the fourth edition, and the chapters that it contained have been relocated to the most relevant sections of the text. Vision and neurotransmission have been moved to part 6, and immunobiology and carcinogenesis, along with a new chapter on AIDS and the HIV virus, have been moved to part 7, "Storage and Utilization of Genetic Information."

Detailed Content Changes of the Fourth Edition

Part 1, "An Overview of Biochemistry and Bioenergetics," contains changes only to part 1, with the addition of the new section entitled "The First Living Systems Were Acellular." This section contains a brief but authoritative account of the most significant prebiotic events that are believed to have lead to the origin of life.

Part 2, "Protein Structure and Function," starts with a new chapter, "The Structure and Function of Water." This chapter pulls together information that was previously scattered in several early chapters and adds new information on the general nature of buffers and the physiologically important buffers containing either phosphate or carbonate. In chapter 4 some information on the determination of amino acid composition has been added. Material in chapter 5 on the three-dimensional structures of proteins has been substantially revised in the middle and final parts. In particular, the discussion of globular protein structures has been reorganized. The structural motif is introduced as the fundamental unit of tertiary structure. Then it is shown how domains are built from single motifs or combinations of motifs and so on. Sections on nuclear magnetic resonance and optical rotatory dispersion and circular dichroism have been added to the last part of chapter 5.

In chapter 6 the discussion of the way in which various factors negatively affect oxygen binding by hemoglobin has been elaborated upon. Thanks to new structural information on myosin (see the references), it has been possible to describe the actin-myosin cycle associated with muscular contraction in greater molecular detail (see figures 6.19, 6.20, and 6.21). In box 6A a comprehensive explanation of multiple binidng and the Hill plot is given. In box 6B a detailed explanation of the inheritance pattern of genetic defects is presented that will be useful for the discussion of hemoglobin in this chapter and will continue to be useful at many points throughout the text. Chapter 7 is a new chapter in which the methods of characterization and purification of proteins are presented. In the third edition this discussion was tacked onto the previous chapter, which dealt with the functional diversity of proteins. Only minor changes have been made to the content of this chapter.

Part 3 begins with chapter 8, of which substantial portions of the middle sections have been rewritten in the interest of achieving greater clarity. Some complex equations have been eliminated without any substantial loss of substance. The section entitled "The Henri-Michaelis-Menten Treatment Assumes That the Enzyme-Substrate Complex Is in Equilibrium with Free Enzyme and Substrate" is new.

The changes in chapter 9 are probably the most important changes made in this text. This is due to a new development in the field of enzymology that invites a reconsideration of many enzyme mechanisms that have been proposed. The second main section of chapter 9, which is concerned with detailed mechanisms of enzyme catalysis, has been substantially revised. In two cases, trypsin and triose phosphate isomerase, the revised mechanisms have resulted from the realization that extra-strong hydrogen bonds are probably formed in the activated complexes. Extra-strong hydrogen bonds known as low-barrier hydrogen bonds were identified in the early 1970s, but their significance in the field of enzyme catalysis has come to light only recently, partly as a result of papers published by W.W. Cleland and Kreeway and Perry Frey and his co-workers (see references and box 9B). Revision of the proposed mechanism for RNase A action has resulted from a paper by Ron Breslow. Several new computer-generated figures have been added to aid in visualization of the overall enzyme structure and the active site (see figures 9.6, 9.7, 9.25, and 9.26). Box 9D discusses catalytic antibodies.

Chapter 10 contains the description of how the mechanism of action of calmodulin has been significantly amplified and updated as a result of new structural information. In chapter 11 the section on coenzyme A has been expanded. The role of ascorbic acids in maintaining the enzyme that catalyzes hydroxyproline formation in collagen is explained in a new short section. Finally, a new box 11A discusses aconitase as an example of an enzyme containing an iron-sulfur complex that is involved in something other than an oxidation-reduction reaction.

In chapter 12 a new section, "A Regulated Reaction Is Effective Only if It Is Exergonic," was added to give emphasis to a principle that was made clear earlier in the text. Although chapter 13 is a new chapter, the material in it is not new; it comes from the first part of chapter 6 in the third edition, which deals with the structures and functions of simple sugars and polysaccharides.

As seen in chapter 14, it is quite remarkable how new understandings of the ancient subject of carbohydrate metabolism keep appearing as we learn more and more about the enzymes that are involved. An expanded discussion of the hexokinases found in different tissues explains why glucose is normally absorbed by most tissues but not by the liver. The hexokinases in question are assembled from different isozymes. An expanded discussion of the different isozymal forms of the bifunctional enzyme that regulates the level of fructose-2,6-bisphosphate explains the different responses of liver and muscle tissue to the hormone epinephrine. The metabolism of the two most important dietary disaccharides is presented with particular reference to the problems they can produce when there are metabolic lesions. A new box 14A describes the history of events leading to the discovery of the glycolytic pathway. A new box 14B describes factors that influence the levels of the regulatory molecule glycerate-2,3-bisphosphate in erythrocytes. This subject is taken up here because the compound in question is also an intermediate in the glycolytic pathway. Chapter 15 contains only minor changes.

The headings in chapter 16 are similar to the ones in the comparable chapter (15) of the third edition. Despite this, many sections have undergone considerable revision. The proton translocation that is powered by electron transport is described as resulting from either redox loops or proton pumps. There are still considerable mysteries concerning how these processes work. Recent crystal structure studies of the F_1 subunit of the ATP-synthase have elevated our understanding of the mechanism for ATP synthesis.

In addition to many minor changes, new information is presented in chapter 17 on the operation of the antenna systems at the photocenters and the mechanism of oxygen evolution. Chapter 18 is derived from chapter 21 of the third edition. In addition, it includes a description of the relevant structures that were presented in chapter 6 of the third edition. Except for reorganizational changes, there have been no major additions or deletions within the contents of the chapter.

Chapter 19 on lipids and membranes, the beginning of part 5, has been completely rewritten to yield a more exciting and shorter presentation. Chapter 20 on mechanisms of membrane transport has been completely rewritten, relocated, and shortened. This subject was relocated because it was considered highly desirable to present the mechanisms of transport before getting too far into the metabolism section. Although the chapter has been considerably shortened, I do not feel that this has resulted in a serious loss of substance.

In chapter 21, only a few changes have been made to accommodate recent advances. In chapter 22, many small changes have been made in the coverage of biosynthesis of membrane lipids. These changes were made to update the material, but clarity of explanation was also addressed. Chapter 23 on cholesterol metabolism includes new information on the factors that control the rate of mevalonate synthesis; the molecular basis of the inherited disease abetalipoproteinemia in which patients have no chylomicrons, VLDLs, or LDLs in the bloodstream; a description of how cholesterol suppresses the synthesis of LDL receptors; and box 3A, a description of the isoprenylation of proteins.

Chapters 24 and 25 on amino acid metabolism are closely knit, as they were in the third edition. The material has been realigned to make it more accessible and to make it easier to use either chapter independently of the other. Chapter 24 deals with amino acid biosynthesis and nitrogen fixation in plants and microorganisms. Important changes in the regulation of the enzyme glutamine synthase have been added to chapter 24. Chapter 25 deals with amino acid metabolism in vertebrates. I suspect that if time is limited, chapter 25 will be favored because of the somewhat greater interest in this subject in most college courses. An appendix that discusses detailed aspects of many catabolic pathways has been added to chapter 25.

Chapter 26 on nucleotides has been considerably updated to accommodate the rapidly changing subject of inhibitors of nucleotide synthesis and their role in chemotherapy. In chapter 27 several sections have been substantially revised to bring them up to date. These sections are "General Aspects of Cell Signaling," "The Adenylate Cyclase Pathway Is Triggered by a Membrane-Bound Receptor," "Protein Phosphorylation Is the Most Common Way in Which Regulatory Proteins Respond to Hormonal Signals," "Variability in G Proteins Adds to the Variability of the Hormone-Triggered Response," "Guanylyl Cyclase Can Be Activated by a Gas," and "Growth Factors Are Proteins That Behave Like Hormones."

In chapter 28 on neurotransmission the section entitled "The Acetylcholine Receptor Is the Best-Understood Neurotransmitter" has been revised to bring it up to date. Two new sections have been added: "Synaptic Receptors Coupled to G Proteins Produce Slow Synaptic Responses" and "Synaptic Plasticity and Learning." Minor changes have been made in chapter 29 on vision to bring the subject up to date.

In chapter 30 a new section, "Helical Structures That Use Additional Kinds of Hydrogen Bonding," deals with the conformational variants formed by telomeres. Four new sections have been added to chapter 31: "Initiation of Chromosomal Replication in Eukaryotes," "The Mismatch Repair System Is Important for Maintaining Genetic Stability," "Some Transposable Genetic Elements Encode a Reverse Transcriptase," and "Bacterial Reverse Transcriptase Catalyzes Synthesis of a DNA-RNA Molecule." Other sections ("SV40 Is Similar to Its Host in Mode of Replication," "Several Systems Exist for DNA Repair," and "Telomerase Facilitates Replication at the Ends of Eukaryotic Chromosomes") have been significantly modified. These additions reflect important advances in our understanding of chromosome replication in eukaryotic systems and DNA repair in both prokaryotic and eukaryotic systems.

For chapter 32 there has been an enormous amount of activity in this field but surprisingly few fundamental advances. This is particularly true in the field of application to practical problems (plenty of sizzle but no steak). Two new sections have been added: "Yeast Artificial Chromosomes (YACS) Are Used for Cloning Fragments as Large as 500 kb in Length" and "Will Nucleic Acids Become Useful Therapeutic Agents?"

In chapter 33, two new sections have been added: "Comparison of *E. coli* RNA Polymerase with DNA PolI and PolIII" and "RNA Editing Involves Changing of the Primary Sequence of a Nascent Transcript." Several existing sections have been updated. The most important updated section, "Messenger RNA Transcription by Polymerase II," has resulted from genetic and biochemical studies of yeast RNA polymerase. Many heretofore unrecognized proteins have been unequivocally recognized as key components of RNA polymerase II. The popular model for stepwise assembly of the RNA polymerase II-DNA complex must now share the limelight with a model in which the holoenzyme remains largely intact in its association-dissociation cycle at the promoter.

In chapter 34, three new sections have been added: "In Addition to the P Site and the A Site for Binding tRNAs, the Ribosome May Possess a Third Site, the E Site," "Protein Folding Is Mediated by Protein Chaperones," and "ATP Plays Multiple Roles in Protein Degradation." Ten sections have been updated: "Ribosomes Are the Site of Protein Synthesis," "The Code Is Highly Degenerate," "Each Synthase Recognizes a Specific Amino Acid and Specific Regions on Its Cognate tRNA," "Aminoacyl-tRNA Synthases Can Correct Acylation Errors," "Translation Begins with the Binding of mRNA to the Ribosome," "Three Elongation Reactions Are Repeated with the Incorporation of Each Amino Acid," "Two (or Three) GTPs Are Required for Each Step in Elongation," "Targeting and Posttranslational Modification of Proteins," "Proteins Are Targeted to Their Destination by Signal Sequences," and "Ubiquitin Tags Proteins for Proteolysis."

Three new sections have been added to chapter 35: "Helix-Turn-Helix Regulatory Proteins Are Symmetrical," "DNA-Protein Cocrystals Reveal Gross Features of the Complex," and "RNA Can Function as a Repressor." The first two of these new sections reflect the progress that has been made in structural studies of DNA regulatory protein complexes. The third relates to the increasing number of examples in which DNA functions as a transcription regulator.

Regarding chapter 36, enormous progress has been made in the area of gene-regulatory proteins in eukaryotes. This has resulted in several new sections and several modified sections in the central part of this chapter. In addition to this, the section entitled "DNA Methylation Is Correlated with Inactivated Chromatin" has been updated by the finding of a long-sought-for genetic correlate. A long-overdue section on how translation controls transcription in eukaryotes has been added. The chapter ends with a new section entitled "Early Development in *Drosophila* and Vertebrates Shown Striking Similarities."

Two new sections have been added to chapter 37. The first, "T-Cell Action Is Frequently Augmented by the Secretion of Hormonelike Proteins Called Interleukins," deals with the hormonelike proteins that bias the interactions between cells of the immune system. The second, "The Immune System in Action," describes three examples of how the immune system disposes of specific invaders by choosing the most appropriate weapons from its arsenal.

Four new sections have been added to chapter 38, and another four have been substantially revised to bring them up to date. These sections comprise most of the latter half of this chapter. The changes help to delineate the distinction between protooncogenes and tumor suppressor genes. In the new chapter 39 the causes and progression of the disease known as AIDS are discussed at an introductory level. Approaches for preventing the spread of AIDS and treating AIDS patients are also discussed.

Suggestions for Teaching and Exams

Let me start out by describing my own teaching experiences and plans. I will be teaching the first semester of biochemistry for the 1997-98 year to Columbia undergraduates. Each week there will be 150 minutes of lectures and 1-2 hours of recitation. This is part of a two semester course. Normally about 140 students register in the first semester; this drops to about half in the second semester. In the first semester I cover the first six parts of this text; in the second semester we cover part 7. It goes without saying that the first semester is the hardest part of this course. Furthermore the subject matter that is covered in the first semester, especially the catalysis and metabolism sections are not encountered to any significant extent in any other departmental course.

This text contains substantially more material than I could cover in one semester. That is intentional and gives users many options; it also gives the student additional material that he/she could review at his/her leisure, perhaps in the following summer.

In my course I will attempt to cover most of the contents of twenty-one out of the first twenty-nine chapters. This will include chapters 2-6, 8-17, 19-21, and 25-27. In chapters 9 and 10 I will probably skimp on the number of specific enzymes that I cover. For example, in chapter 10 I will probably only cover phosphofructose kinase in the latter half of the chapter, which discusses specific enzymes in some detail. In chapter 11, which deals with coenzymes, I will only cover the first thirteen pages through folate coenzymes. Depending on how things go I may do some cutting in chapters 16 and 17. In chapter 21 I will cut out material on unsaturated fatty acids. And finally in chapter 26 I will do some cutting.

Other experienced teachers of this subject may have the lectures already worked out and will just assign the most relevant sections as the course progresses. The following are a few pieces of advice that I would like to offer:

1. Don't try to cover too much. This is a variable that depends upon the caliber of students you are teaching.

2. A lecturer who follows the text rather closely is the easiest to follow.

3. It is perfectly all right if you do more cutting than I do. It is far better to cover a small number of subjects thoroughly than to skim over many superficially.

4. It is very important for students to attack the work study problems. There are different ways in which you might persuade them to do this. Keep in mind that the students will be most highly motivated to do the work study problems if they think they are related to the exams. What I do is to make up my exams well ahead of time so that I can emphasize related matters in class and recommend relevant work study problems.'

5. The bulk of the questions in the test item file are multiple choice to ease in the grading of large classes and minimize bickering. I give a short practice test before the first exam that counts. The students really appreciate this. I like to give open book tests but in recent years I have shied away from this because I find that most students do not study as hard as they should.

There are so many facts in biochemistry that the students are bound to ask you (many times) what they have to know for the exam. This is a hard question to answer. I usually answer "You should know such and such a pathway cold, but in the case of another pathway, it is sufficient if you get the overall gist of what's going on." For example, if we were talking about chapters 14 and 15, which deal with glycolysis and the citric acid cycle, I would probably say that they should know the glycolytic pathway and the TCA cycle cold and other reactions in the chapter including the pentose phosphate pathway should be understood only for overall function. Incidentally, I devote at least three lectures to chapter 14 and two lectures to chapter 15. These are the first metabolism chapters and probably contain more material to master than the other chapters on metabolism.

To avoid overtaxing the student's memory capacity, I give four equally weighted exams and the students are only responsible for one-fourth of the course material for each exam. This means I have done away with the traditional final.

SOLUTIONS MANUAL

Prepared by

Larry Loomis-Price
Gwen Shafer

Boston   Burr Ridge, IL   Dubuque, IA   Madison, WI   New York   San Francisco   St. Louis
Bangkok   Bogotá   Caracas   Lisbon   London   Madrid
Mexico City   Milan   New Delhi   Seoul   Singapore   Sydney   Taipei   Toronto

CONTENTS

2 Thermodynamics in Biochemistry

Chapter 2 Answers

2. State function describes the thermodynamic parameters of the system under consideration at a particular moment. Changes in those parameters change the thermodynamic state of the system being considered. Only the difference between initial and final states, not the path taken to achieve these states, is important in most thermodynamic considerations. Enthalpic contributions defining the thermodynamic state are considered only at the initial and final states. Enthalpy is independent of pathway and is therefore a state function.

4. Consider that $\Delta H = \Delta(E + Pv)$

   $$\Delta H = \Delta q - \Delta w + P\Delta v + v\Delta P$$

   At constant pressure and minuscule volume changes associated with most biochemical reactions, the terms Δw, $P\Delta V$, and $V\Delta P$ may be considered zero. Thus, $\Delta H = \Delta q$ and enthalpy (H) is equal to heat gain or loss at constant pressure (q_P).

6. ATP has two phosphoanhydride bonds, each of which releases 7.5 – 12 kcal/mole upon hydrolysis. The high free energy of hydrolysis may be attributed in part to:

   (a) decreased negative charge density in the products compared to the triphosphate portion of ATP;

   (b) increased resonance stabilization of the products compared to ATP. (See discussion in text.)

   Cleavage of the phosphoanhydride bonds of ATP may be coupled to biochemical reactions:

   (a) ATP → ADP + P_i

   (b) ATP → AMP + pyrophosphate

   The standard state free energy of hydrolysis of the phosphoanhydride bonds occupies a position in the middle of free energies of hydrolysis of organophosphate compounds that are encountered in metabolism. Mixed phosphoric anhydrides (e.g., glycerate-1,3-bisphosphate), phosphoenolates (phosphoenolpyruvate), and phosphoramides (phosphocreatine) may be used to phosphorylate ADP. ATP is consumed in phosphoryl or adenyl transfer reactions. ATP thus may be considered a "high energy phosphate" transfer agent central to metabolism.

6. (continued)

From thermodynamic considerations, the hydrolysis of ATP,

(1) $ATP + H_2O \rightarrow ADP + P_i + H^+$ $\Delta G^{\circ\prime} -7.5$ kcal/mole

may be occupied to the phosphorylation of a compound, for example, glucose.

(2) glucose + P_i → glucose-6-phosphate H_2O $\Delta G^{\circ\prime} + 3.3$ kcal/mole

Each of the reactions (1 and 2) may be considered a separate partial reaction of

glucose + ATP → glucose-6-phosphate + ADP $\Delta G^{\circ\prime} -4.5$ kcal/mole

However, neither ATP nor other phosphorylated biological molecules undergo hydrolysis to free phosphate, with the free phosphate being in turn added to an acceptor. Specific enzymes are required to decrease the activation energy of phosphoryl transfer. From a kinetic consideration, the phosphate from ATP is transferred directly to the substrate in the enzyme active site or through an enzyme-bound intermediate during catalysis. In either of the processes described, the final thermodynamic states are identical but the kinetic considerations differ markedly.

8. a. The standard state free energy change can be calculated from the equilibrium constant,

$$\Delta G^{\circ\prime} = -2.3RT \log K'_{eq}$$

$$\Delta G^{\circ\prime} = -1.36 \text{ kcal/mole (log 59.5)}$$

$$\Delta G^{\circ\prime} = -2.4 \text{ kcal/mole}$$

b. The hydrolysis of ATP can be determined by the combination of reactions (P3) and (P4) as follows:

(1)   Creatine phosphate + ADP + H^+ → ATP + creatine + $\Delta G^{\circ\prime} = -2.4$ kcal/mole (P3)

(2)   Creatine + P_i → creatine phosphate + H_2O $\Delta G^{\circ\prime} = +10.5$ kcal/mole (P4)

(3) $H^+ + ADP + Pi \rightarrow ATP + H_2O$ $\Delta G^{\circ\prime} = +7.9$ kcal/mole (P5)

Therefore, hydrolysis is the reverse of reaction (P5).

ATP + H_2O → ADP + Pi + H^+ $\Delta G^{\circ\prime} = -7.9$ kcal/mole

10. a. Glycerate-1,3-bisphosphate H_2O → glycerate-3-phosphate + P_i ($\Delta G^{\circ\prime} = -11.8$ kcal/mole)

Creatine + P_i → phosphocreatine + H_2O ($\Delta G^{\circ\prime} = +10.3$ kcal/mole)

Glycerate-1,3-bisphosphate + creatine → phosphocreatine + glycerate-3-phosphate
($\Delta G^{\circ\prime} = -1.5$ kcal/mole)

Thermodynamically favorable as written.

10. (continued)

  b. $\text{D-glucose-6-P}_i + \text{H}_2\text{O} \rightarrow \text{D-glucose} + \text{P}_i$ $\Delta G^{\circ\prime} = -3.3$ kcal/mole

  $\text{D-glucose} + \text{P}_i \rightarrow \text{D-glucose-1-P}_i + \text{H}_2\text{O}$ $\Delta G^{\circ\prime} = +5.0$ kcal/mole

  $\text{D-glucose-6-P}_i \rightarrow \text{D-glucose-1-P}_i$ $\Delta G^{\circ\prime} = +1.7$ kcal/mole

Thermodynamically unfavorable as written.

(The phosphate linkage to the 6-OH group of glucose is a phosphate ester (a phosphorylated alcohol). The phosphate linkage to the C-1 position is a phosphorylated hemiacetal OH, thus the difference in free energy of hydrolysis of the two compounds.)

  c. $\text{Phosphoenolpyruvate (PEP)} + \text{H}_2\text{O} \rightarrow \text{pyruvate} + \text{P}_i$ $\Delta G^{\circ\prime} = -14.8$ kcal/mole

  $\text{H}^+ + \text{ADP} + \text{P}_i \rightarrow \text{ATP} + \text{H}_2\text{O}$ $\Delta G^{\circ\prime} = +7.5$ kcal/mole

  $\text{H}^+ + \text{PEP} + \text{ADP} \rightarrow \text{ATP} + \text{pyruvate}$ $\Delta G^{\circ\prime} = -7.3$ kcal/mole

Thermodynamically favorable as written.

  d. $\text{Glycerol-3-phosphate} + \text{H}_2\text{O} \rightarrow \text{glycerol} + \text{P}_i$ $\Delta G^{\circ\prime} = -2.2$ kcal/mole

  $\text{H}^+ + \text{ADP} + \text{P}_i \rightarrow \text{ATP} + \text{H}_2\text{O}$ $\Delta G^{\circ\prime} = +7.5$ kcal/mole

  $\text{H}^+ + \text{Glycerol-3-phosphate} + \text{ADP} \rightarrow \text{ATP} + \text{glycerol}$ $\Delta G^{\circ\prime} = +5.3$ kcal/mole

12.

| $\dfrac{2500 \text{ cal}}{\text{day}}$ | $\dfrac{1 \text{ g glc}}{4 \text{ cal}}$ | $\dfrac{1 \text{ mol glc}}{180 \text{ g glc}}$ | $\dfrac{30 \text{ mol ATP}}{1 \text{ mol glc}}$ | $\dfrac{660 \text{ g ATP}}{1 \text{ mol ATP}}$ | $\dfrac{1 \text{ lb}}{454 \text{ g}}$ | $= 151$ lbs/day |

The mass portion of this figure is comparable to the weight of the individual!

14. The energy available in citryl-CoA could easily be captured in an ATP by a mechanism analogous to succinyl-CoA synthetase. However, because the reaction catalyzed by malate dehydrogenase (reaction 18) is so unfavorable and is "pulled" by the equilibrium position of the next enzyme (reaction 17), any change could be critical. The result would be two sequential reactions having very low product concentrations which in turn could have a significant impact on metabolism.

3 The Structure and Function of Water

Chapter 3 Answers

2. Insolubility reflects a more significant collection of interactions between solute molecules than between solute and solvent molecules. A number of amide-amide hydrogen bonds are possible that can form ring or chain structures. Amides have a significant resonance form in which the nitrogen and oxygen carry positive and negative charges respectively. This resonance allows amides to interact with each other as dipoles. The increased polarization also substantially increases the strength of the amide-amide hydrogen bonds.

4. $3.4 \times 10^{-4.87} = 1.34 \times 10^{-5} = Ka = \dfrac{[H^+][Pr^-]}{[HPr]}$

   $[HPr] + [Pr^-] = 0.012\ M$ $X = [H^+] = [Pr^-]$

   $1.34 \times 10^{-5} = \dfrac{X^2}{0.012 - X}$

   $0 = X^2 + 1.34 \times 10^{-5}X - 1.61 \times 10^{-7}$

   positive solution via the quadratic equation: $X = 3.95 \times 10^{-4}$

   or pH = 3.4

6. Procedure (a) will produce the correct buffer but it will also contain an additional salt (KCl). Buffer (b) will not be of the correct molarity. The H_3PO_4 added to reach the needed pH will increase the concentration of the buffer. Note that the buffer concentration (in this situation) is the sum of the concentrations of the species containing the phosphate moiety. Buffer (c) may have the needed pH and concentration but it depends upon the purity of both salts and relies on a pKa value that may not be correct. Method (d) is the normal method used to prepare a buffer. Why wasn't the cheaper NaOH used to adjust the pH in buffer (d)?

8.   Initial Situation   Final Situation

$6.80 = 6.39 + \log \dfrac{[cit^{3-}]}{[HCit^{2-}]}$ $6.50 = 6.39 + \log \dfrac{[cit^{3-}]}{[HCit^{2-}]}$

$[cit^{3-}] + [HCit^{2-}] = 0.150$ M

$2.57 = [cit^{3-}]/[HCit^{2-}]$ $1.29 = [cit^{3-}]/[HCit^{2-}]$

$[HCit^{2-}] = 0.042$ M $[HCit^{2-}] = 0.066$ M

The increase in [HCit2-] is 0.024 M

$$\dfrac{10 \text{ mL}}{} \left| \dfrac{1 \text{ L}}{1000 \text{ mL}} \right| \dfrac{0.024 \text{ mol}}{\text{L}} \left| \dfrac{1000 \text{ millimol}}{1 \text{ mol}} \right. = 0.24 \text{ mmol HCl}$$

10. $(11.7 \text{ g KH}_2\text{PO}_4/0.5 \text{ L}) \times (1 \text{ mol}/136.1\text{g}) = 0.172$ M

Using the Henderson-Hasselbalch equation:

$$7.10 = 7.21 + \log \dfrac{[HPO_4{}^{2-}]}{[H_2PO_4{}^-]}$$

and

$$[HPO_4{}^{2-}] + [H_2PO_4{}^-] = 0.172 \text{ M}$$

gives

$$[H_2PO_4{}^-] = 0.0968 \text{ M}, \qquad [HPO_4{}^{2-}] = 0.0752 \text{ M}$$

and

$$7.10 = 2.12 + \log \dfrac{[0.0968M]}{[H_3PO_4]}$$

$$[H_3PO_4] = 1 \times 10^{-6} \text{ M}$$

12.   2.45 g H_3PO_4/97 g/mol = 0.025 mol H_3PO_4

2.45 g KOH/56.1 g/mol = 0.044 mol KOH

after reactions: 0.006 mol $H_2PO_4{}^-$

0.019 mol $HPO_4{}^{2-}$

$$pH = 7.21 + \log \dfrac{[0.019 \text{ mol } HPO_4{}^{2-}/600 \text{ mL}]}{[0.006 \text{ mol } H_2PO_4{}^-/600 \text{ mL}]} = 7.71$$

concentration: H_3PO_4 provided all of the phosphate thus

0.025 mol/0.6 L = 0.042 M

4   The Building Blocks of Proteins: Amino Acids, Peptides, and Polypeptides

Chapter 4 Answers

2. The isoelectric point of an amino acid (or protein) is the pH at which the sum of positive and negative charges is zero. Fully protonated histidine will bear a net +2 charge. (α-amino and imidazole group are protonated, each with a positive charge. The α-carboxylate is protonated but is neutral.)

   There are two approaches to the problem. By inspection, we note that removal of two protons will decrease the net charge to zero. Thus, 0.051×10 mmoles/l = 0.5 mmoles of histidine. Two mmoles of OH^- are required to titrate each mmole of hisitidine. Thus 1 mmole of NaOH must be added.

   The second approach requires quantifying the final state of protonation of histidine at the isoelectric point. The pl is $(pK_a + pK_R)/2 = 7.69$. At that pH, the α-carboxyl (pK 1.8) will be fully deprotonated and the deprotonation is accomplished by the addition of 0.5 mmole of NaOH to the fully protonated molecule.

   Apply the Henderson-Hasselbach equation to determine the ratio of unprotonated to protonated imidazole (R) group.

   $$7.69 = 6.0 + \log \frac{(Im)}{(ImH^+)}$$

   $$\log \frac{(Im)}{(ImH^+)} = 1.69$$

   $$\frac{(Im)}{(ImH^+)} = 49; \text{ but } (Im) + (ImH^+) = 0.5 \text{ mmoles, so}$$

   $$(Im) = 0.49 \text{ mmoles}$$

Thus, 0.49 mmoles of NaOH are required to titrate the imidazole to the appropriate ratio of unprotonated to protonated form. By a similar calculation, it can be shown that the amount of unprotonated (α-NH$_2$) is 0.01 mmoles, requiring the addition of 0.01 mmoles of NaOH. The total NaOH required is thus $(0.5 + 0.49 + 0.01)$ mmoles or 1 mmole NaOH.

Note that when the pH is below the pK_a, the group is more protonated (neutral or positively charged) than unprotonated, but when the pH is above pK_a, the group is more unprotonated (neutral or negatively charged) than protonated.

4. The ionizable groups will be more protonated if pH is less than pK_a and will be more unprotonated if pH is greater than pK_a. Examine the table of amino acid pK_a values in the text and determine (a) which of the ionizable R-groups will be protonated at each pH value and (b) whether protonation neutralizes a negative charge or confers a positive charge. Those whose side chains are charged are

    (a) pH 2: Arg, His, Lys

    (b) pH 7: Arg, Asp, Glu, Lys

    (c) pH 12: Arg, Asp, Glu, Tyr, Cys

6.

$$0.5 \ mM = [A] + [B]$$

$$12.18 = 12.48 + \log \frac{[B]}{[A]}$$

$$0.3 = + \log \frac{[B]}{[A]}$$

$$2 = \frac{[B]}{[A]}$$

$$0.17 \ mM = [A]$$

$$0.33 \ mM = [B]$$

$$\text{Charge} = \frac{(0.17 \ mM)(0) + (0.33 \ mM)(-1)}{0.17 \ mM + 0.33 \ mM} = +0.66$$

8. Rose was focusing on the geometry of carbon three of threonine in his comparison with the two sugars. He found the third carbon of threonine to match the geometry of the third carbon of D-threose and unfortunately titled his paper "D-threonine" (naming the amino acid and its geometry for the sugar). The geometry of the highest numbered chiral carbon in a sugar determines the D/L nature; this is carbon number three in the case of threose. Carbon number two of threose (but not erythrose) was found to have the "same" geometry as carbon 2 of threonine, but it is carbon two that determines the D/L geometry of an amino acid. An editor or reviewer of that article should have prevented confusion by omitting the "D" in D-threonine.

10. The amino acids will be separated based on molecular charge. Consider the charge of each amino acid at pH 6.1. The isoelectric points of the amino acids in the mixture are: glutamic acid, 3.1; alanine, 6.1; and arginine, 10.74.

Amino acid	pI	net charge at pH 6.1
Glutamic acid	3.1	negative (pH > pI)
Alanine	6.1	neutral (pH = pI)
Arginine	10.74	positive (pH < pI)

At pH 6.1, the negatively charged glutamate will bind to the positively charged, weakly basic resin, but arginine and alanine will pass unretarded through the column. A weakly acid ion exchange resin at pH 6.1, bearing a negative charge, will retain the positively charged arginine (0 net charge) and glutamate (net negative charge) will pass unretarded through the column.

A strategy to separate the three amino acids involves operating the columns in series. Each column is buffered at pH 6.1, eliminating complications due to buffer or pH changes. The effluent from the weakly basic ion-exchange column contains arginine and alanine, but arginine will be retained by the acidic column and alanine will be eluted. Washing each column independently with buffer containing increased salt concentration will elute the bound amino acid.

12. Trypsin:

$$(Phe, Val, Ala) Lys_4 \xrightarrow{T_2} (Asp, Gly, Tyr) \xrightarrow{T_1}$$

Trypsin cleaves after Lys

$$(Val, Ala) Phe_3Lys_4Gly_5Tyr_6Asp_7 \xrightarrow{CT_2} \xrightarrow{CT_1} \xrightarrow{CT_3}$$

Chymotrypsin cleaves after Phe and Tyr

$$Ala_1Val_2Phe_3Lys_4Gly_5Tyr_6Asp_7$$

N-terminus = Ala via DNP analysis

DNP-Lys is epsilon-dinitrophenyllysine

14. The peptides could be detected with ninhydrin. Peptides CT_1 and CT_2 have a UV absorbtion because of their aromatic amino acid content and could be detected by their UV absorption. The third chymotrypsin peptide could not be detected by UV.

16.

5 The Three-Dimensional Structures of Proteins

Chapter 5 Answers

2. Both structures are supported by hydrogen bonds between backbone amides. In the alpha helix the hydrogen bonds are between amides that are three amides away on the polymer and are approximately parallel with the long axis of the helix. The hydrogen bonds in the pleated sheet structure are between amides that are quite distant from each other and may be on separate peptide chains. The peptide strands in the sheet structures can be in the same or opposite directions and are in essentially their maximum extension.

4. In order for amino acyl groups to exist in non-permitted regions of the Ramachandran Plot atoms must be compressed. The energy required to compress these atoms must come continually from other interactions (hydrophobic interactions, hydrogen bonds, etc.) elsewhere within the protein.

6. In principle, proteins should assume a conformation yielding the lowest free-energy level. Entropic and enthalpic changes in the system (peptide plus surrounding medium) should sum to a negative value for the folding of the protein. Recall that $\Delta G = \Delta H - T \Delta S$. Thus, interactions that decresae ΔH or increase ΔS contribute to a more negative ΔG. Organized structure in the protein results in a decrease in entropy of the protein. This decrease in entropy must be offset by an increase in the surroundings. Removing hydrophobic residues from the aqueous interface is thermodynamically favorable and accounts, in part, for the increased entropy in solution surrounding the protein. Water molecules are more highly organized in the space immediately surrounding the hydrophobic residues than in bulk water. Shifting the hydrophobic residues from an aqueous to an anhydrous environment decreases organization of the surrounding water and increases the entropic contributions to folding. In globular proteins dissolved in aqueous media, one finds that in general the hydrophobic residues are exposed on the surface of the protein to interact with water, whereas the hydrophobic residues coalesce in the core or interior of the protein. However, as noted in the text, enthalpically favorable interactions offset the decreased entropy of the folded protein and contributes to stabilization of the folded protein.

8. Surface or solvent exposed residues hydrogen bond with water molecules and are important in solubility of the protein in aqueous media. Replacement of a surface hydrophilic with a hydrophobic residue would increase the organization of water molecules around the hydrophobic residue and would be expected to decrease, by a small amount, the overall structure stability of the protein molecule as well as the surrounding solvent. Changing an interior hydrophobic to a hydrophilic residue likely also would destabilize the structure by potentially disrupting an area of hydrophobic interaction. The hydrophilic character of the residue, with or without a formal charge, would preclude close approach of a neighboring hydrophobic residue. Each of these replacements would diminish, to a small degree, the overall structural stability of the protein. Single substitutions would likely be accommodated without major disruption to the overall structure of the protein, but there are exceptions. Substitution of valine for glutamic acid at position 6 of the β subunit of hemoglobin results in aggregation of the deoxyhemoglobin (sickle-cell hemoglobin). Erythrocyte flexibility is markedly diminished.

10. Replacement of Leu with Asp would not necessarily alter the secondary structure; α-helix formation is not precluded. However, the helix containing the aspartate residues would be considerably more hydrophilic and would be more difficult to insert into the membrane. It would be energetically less favorable to remove an anionic residue from the aqueous environment, where the charge is somewhat shielded by water, and insert the charge into a nonpolar environment. The peptide in which 50% of Leu was replaced with Asp would not be likely to insert into the membrane.

12. The protein (200 K M_r) is dissociated into two units of identical molecular weight (100 K M_r) when denatured with guanidinium chloride in the absence of reductant. Thus, these 100K M_r units are not linked by disulfide bonds. However, each 100 K M_r unit is comprised of two subunits (75 K and 25 K) that are linked by disulfide bonds. Treatment of the protein with the denaturant under reducing conditions yields 75 K and 25 K M_r proteins. The structure can be proposed schematically:

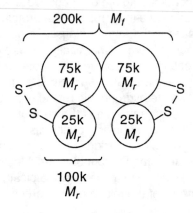

14. Val, Ile, Leu, Ala, Phe, i.e., hydrophobic R groups.

6 Functional Diversity of Proteins

Chapter 6 Answers

2. $H_2CO_3 \rightleftarrows HCO_3^- + H^+$

$$K_{eq} = \frac{[HCO_3^-][H^+]}{[H_2CO_3]}$$

$pK_{eq} = 6.4 \quad K_{eq} = 10^{-6.4} = 3.98 \times 10^{-7}$

$[H^+] = 10^{-7.4} = 3.98 \times 10^{-8}$ M

$$3.98 \times 10^{-7} = \frac{[HCO_3^-][3.98 \times 10^{-8}]}{[H_2CO_3]}$$

$$10 = \frac{[HCO_3^-]}{[H_2CO_3]} = \frac{10}{1} \quad \text{or} \quad \frac{(10)(100)}{(10+1)} = 91\% \text{ ionized}$$

4. The most convenient means used to detect abnormal hemoglobins is by electrophoresis. This is easily done by diluting blood with distilled water and using the resulting red blood lysate. In addition, because of its color, the migration of hemoglobins is readily visualized. Separate by electrophoresis is based primarily on charge; any sample that migrates to a position different from that of "normal;" hemoglobin, most likely has a charge difference from "normal" hemoglobin. The simplest source of a charge difference is an altered amino acid composition.

6. Adult hemoglobin has to have an affinity sufficient to absorb (bind) oxygen from that dissolved in the lung tissues. Fetal hemoglobin must compete with adult hemoglobin across the placental barrier. For any net transfer of oxygen from the mother to the fetus, fetal hemoglobin must have a greater affinity for oxygen than adult hemoglobin.

8. In order for a recognition of chirality there must be a "three point" interaction. For example, if only the carboxylate and the 3-phospho-group on glycerate-2,3-bisphosphate were recognized then it would be anticipated that both the *D*- and *L*- forms of glycerate-2,3-bisphosphate would be effective in regulating the affinity of hemoglobin for oxygen. However, if a three (or four) point recognition occurs then the interaction can be specific for one chiral form. Notice in Figure 6.9 that the interaction between glycerate-2,3-bisphosphate involves several (ca. 7) positive charges on the hemoglobin. The interaction is specific for *D*-glycerate-2,3-bisphosphate.

10. Some of the CO_2 is carried to the lungs as dissolved CO_2, bicarbonate, carbonate or carbonic acid. Also, only the oxidation of carbohydrates require as many moles of O_2 as CO_2 produced. Fats require more moles of O_2 than CO_2 produced; therefore, more O_2 than CO_2 is carried by hemoglobin.

12.

$$\frac{169\text{min} \times 1/3 \text{ hp}}{} \left| \frac{178 \text{ cal}}{1 \text{ hp} \times 1 \text{ sec}} \right| \frac{60 \text{ sec}}{1 \text{ min}} \left| \frac{1 \text{ Kcal}}{1000 \text{ cal}} \right| \frac{\text{mol ATP}}{7.5 \text{ Kcal} \times 0.5} = 160 \text{ mol of ATP}$$

14. The single feature common to both proline and glycine is the ϕ-ψ angles permitted these amino acids. (see the Ramachandran Plot). Any tight turn typically requires the flexibility of a glycine and/or a proline. An alpha helix "cannot" include a proline and rarely includes a glycine. These amino acids frequently indicate the end of a helix. Glycine has the smallest volume of any amino acid and can be used in interior situations in which another amino acid might be too large.

16. Glycerate-2,3-bisphosphate (GBP) binds in the pocket in the center of the hemoglobin $\alpha_2\beta_2$ tetramer. 2,3-GBP binds more tightly to deoxyhemoglobin than the oxyhemoglobin. In the lung, the O_2 partial pressure is sufficiently large to drive the deoxyHb/oxyHb equilibrium toward oxyHb with subsequent decrease in GBP affinity. In the O_2-consuming tissues, the decreased O_2 partial pressure favors deoxygenation of oxyhemoglobin. The deoxyHb form is stabilized by the GBP, shifting the deoxyHb/oxyHb equilibrium toward the deoxy form. Thus, more O_2 is released to the tissues in the presence of GBP than would occur in its absence. If GBP were not present, or if hemoglobin failed to bind GBP, there would be a diminished supply of O_2 to the tissues and that, if severe, could lead to anoxia and cell death.

7   Methods for Characterization and Purification of Proteins

Chapter 7 Answers

2.   Heat treatment of protein solutions denatures and precipitates some of the proteins, while others remain both soluble and stable. Thermal liability is determined empirically for each enzyme or protein of interest. The enzyme of protein of interest may be rapidly denatured by the thermal treatment, necessitating alternative strategies for isolation.

4.   The ion-exchanger group on the CM-cellulose is a weak carboxylic acid whose pK is around 4. Above pH 4, the carboxylic acid is unprotonated and negatively charged. However, in the buffer system given in problem 3, the protein is negatively charged and will not adhere to the negatively charged CM-cellulose. The protein is positively charged at pH more acidic than the pI. Therefore the protein will probably adhere to (bind to) the CM-cellulose if the pH is between 4 and 6.

6.   Proteins are separated by SIDS-PAGE on the basis of molecular weight of the denatured protein. The molecular weight is the minimum value or subunit molecular weight. Two proteins that differ in some properties but have the same subunit molecular weight will likely appear as a single protein band after SDS-PAGE. Thus, a 40,000 M_r dimer composed of 2 × 20,000 M_r subunits will not be separated from an 80,000 M_r tetramer composed of a 4 × 20,000 M_r subunits. Native, or nondenaturing, electrophoresis separates proteins based on mass/charge characteristics. The student's "pure" protein contains at least two components separable by the criterion of mass/charge but which share a common subunit molecular weight. It is also possible that the multiple protein bands appearing in the nondenaturing gel arose through deamidation of glutamine or asparagine residue side chains.

8. a. DEAE cellulose will have a net positive charge at pH 7 due to protonation of the weakly basic tertiary amine (pK_a approximately 8.5). The exchanger will bind or retard negatively charged proteins. The charge on each protein may be estimated from the pI. At pH values greater than pI, the protein will be negatively charged. At pH below pI, the protein will be positively charged.

Protein	pI	Charge at pH 7
Protein 1	10	(+)
Protein 2	4	(−)
Protein 3	8	(+)
Protein 4	5	(−)

Proteins 1 and 3 should elute in the initial wash buffer, but proteins 2 and 4 are predicted to bind to the column. Based solely on isoelectric point, one might predict that protein 4, then protein 2, would be eluted in the salt gradient.

b. The carboxyl group of the CM-cellulose will be negatively charged at pH 7 because of deprotonation of the weak acid (pK_a approximately 4). The charge on each protein was established in the previous section.

Proteins 2 and 4 would be eluted in the initial wash buffer from the column, whereas proteins 1 and 3 would be predicted to adhere to the column. Based solely on pl values, protein 3 would be predicted to elute prior to protein 1 in the KCl gradient. In each separation of proteins based on ion exchange (e.g., problems 8a and 8b), the actual elution order must be determined experimentally. Proteins having the same or similar pl but differing in the absolute number of charged residues will likely differ in their elution characteristics.

c. Gel exclusion chromatography separates globular proteins on the basis of relative size, which is a function of molecular weight. Larger proteins (higher molecular weight) elute before smaller (lower molecular weight) proteins. The gel exclusion resin proposed to separate the four proteins (8a) has an exclusion limit of 30,000. Proteins with molecular weight greater than the limit (protein 2, 62,000 M_r) are excluded from entry into the gel and elute in the void volume (V_0). The other proteins in the solution will elute in the order protein 3, protein 1, protein 4.

10. The "salting out" procedure can be done on essentially any volume of material. If is often a rapid effective way to reduce the volume of crude extracts and at the same time eliminate a major portion of the total protein. The speeds/simplicity of the process is also important in that it often eliminates proteolytic enzymes which can hydrolyze the desired proteins.

12. There is essentially no difference in the size (mass) of phosphorylase a and phosphorylase b; therefore, they should elute as a single peak upon gel filtration. Phosphoyrlase a has a bound phosphate which would give it a greater negative charge than phosphorylase b at pH's greater than about 2 (pK_{a1} of a phosphate ester). This difference in charge allows the separation of phosphorylase a from phosphorylase b with the use of DEAE-cellulose.

14. The isoelectric point of a protein reflects the ratio of basic to acidic amino acids in that protein. If the ratio of the acidic amino acid residues to basic amino acid residues in a protein is greater than one then the protein will have a higher pl. If the ratio is less than one then the pl will be low. Therefore, the ratio of Arg + Lys + His to Asp + Glu is greater in protein 1 than in protein 2.

16. Proteins, in general, are not soluble in most organic solvents and organic solvents frequently denature proteins. Normally substances crystallize when they are relatively pure. Until the final stages, after several purification procedures, can a protein be called "relatively pure" Proteins can be crystallized but it is not normally thought of as a urification procedure.

18. Two separate molecules of mercaptoethanol are involved in reducing a disulfide bond in a two step process yielding the disulfide form of mercaptoethanol. Only one molecule of DTT is required to reduce a disulfide. The second reaction in the case of DTT is an intramolecular reaction and is favorable in an entropy sense. Also, the formation of the six member dithiane ring is energetically favorable.

8   Enzyme Kinetics

Chapter 8 Answers

2. Let (s_0) be the concentration of substrate present before the reaction is initiated (t_0) and let (S_k) represent the concentration of substrate remaining after time (T_k). After 7 min, 87% of the substrate was consumed and 13% remains. Rearranging the equation for first-order reaction yields

$$2.3 \log \frac{(S_k)}{(S_0)} = -kt_k \text{ and}$$

$$2.3 \log \frac{(S_0)}{(S_k)} = kt_k$$

where $(S_0) = 100\%$ and S_k is 13% after 7 min reaction time.

$$2.3 \log \frac{(100)}{(13)} = k(7 \text{ min})$$

$$k = \frac{2.3 \log (7.7)}{(7 \text{ min})}$$

$$k = 0.29 \text{ min}^{-1}$$

The value of k now may be used to calculate the time required to convert 50% of the substrate to product.

$$2.3 \log \frac{(100)}{(50)} = (0.29 \text{ min}^{-1})(t)$$

$$t = 2.3 \frac{(0.3)}{(0.29)}$$

$$t = 2.4 \text{ min}$$

For a first-order reaction, the time required for conversion of 50% substrate to product is called the half-time $(T_{0.5})$ and is independent of the initial concentration of substrate.

4. The enzyme-substrate complex forms upon binding of substrate at the enzyme-active site, the initial step in catalysis. The substrate at this point is chemically unchanged but is physically restrained and correctly oriented for subsequent reaction. The enzyme-substrate complex may dissociate to regenerate free substrate and enzyme or may continue along the catalytic pathway to an activated enzyme-substrate complex (ES), the transition state. The transition state is the chemical intermediate on the reaction path between substrate and product and is the form present at the highest energy point along the reaction pathway. Although the enzyme-bound transition-state complex is the highest energy intermediate, the energy maximum is lower than that of the uncatalyzed reaction. The transition state intermediate is stabilized by binding to the active site and represents the chemical form of the complex capable of proceeding to product or returning to substrate, with equal probability of either event.

6. a. $K_m = \dfrac{(k_2 + k_3)}{k_1}$

   $K_m = \dfrac{(10^5 \text{ s}^{-1} + 10^2 \text{ s}^{-1})}{(10^9 \text{ M}^{-1} \text{ s}^{-1})}$

   $K_m = 10^{-4} \text{ M}$

   b. $V_m = k_3 (E_T)$ where $E_T = 10^{-10}$ M

   $V_m = 10^{-8} \text{ M s}^{-1}$

   c. Turnover number is moles of substrate converted to product per time unit per mole of enzyme.

   $$\text{Tn} = \frac{V_m}{E_T} = k_3 = k_{cat}$$

   Turnover number = 100 s^{-1}

   d. $v = \dfrac{V_m (S)}{K_m + (S)}$

   Substitute the kinetic constants derived in parts (9a) and (9b) and substrate concentration (2×10^{-5} M).

   $$v = \frac{(10^{-9} \text{ M s}^{-1})(2 \times 10^{-5} \text{ M})}{(10^{-4} \text{ M} + 2 \times 10^{-5} \text{ M})}$$

   $v = 1.72 \times 10^9 \text{ M s}^{-1}$

8. Let $v = \dfrac{V_{max}}{2}$ and substitute into Equation 25.

   $$\frac{V_{max}}{2} = \frac{V_{max} [S]}{[S] + K_M}, \text{ solving gives: } K_M = [S]$$

10. The type of inhibition imposed by each inhibitor can be determined graphically using the Lineweaver-Burk plot. The reciprocal of the velocity measured in the absence and in the presence of each inhibitor is graphed versus the reciprocal of the corresponding substrate concentration.

1/(S) (mM^{-1})	1/Velocity s M^{-1} ($\times 10^{-7}$) Inhibited	1/Velocity s M^{-1} ($\times 10^{-7}$) Inhibitor A	1/Velocity s M^{-1} ($\times 10^{-7}$) Inhibitor B
3.0	0.606	0.952	1.26
2.5	0.538	0.826	0.12
2.0	0.469	0.699	0.980
1.5	0.400	0.575	0.840
1.0	0.334	0.450	0.699
0.5	0.269	0.325	0.559

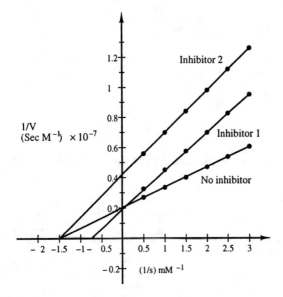

Inhibitor A has an effect on the slope but not on the *y*-intercept of the double reciprocal plot when compared with the data from the unhibited enzyme. Infinite substrate concentration abolishes the effect of inhibitor A and is diagnostic for competitive inhibition. Therefore, inhibitor A is competitive with the substrate for the enzyme active site. The kinetic constants (K_m and V_m) are only defined for the uninhibited enzyme and are derived graphically as shown in problem 11.

V_m is 5.0×10^{-7} M s^{-1} and K_m is 0.67 mM

10. (continued)

The effect of a competitive inhibitor on the velocity of the enzyme-catalyzed reaction is given by

$$v_1 = \frac{V_m(S)}{K_m(1 + I/K_i) + S}$$

The reciprocal of this expression, in Lineweaver-Burk form,

$$\frac{1}{v_1} = \frac{K_m}{V_m}(1 + I/K_i)\frac{(1)}{(S)} + \frac{1}{V_m}$$

reveals that the competitive inhibitor affects the slope term of the kinetic expression but not the Y-intercept term (V_m). The value of K_i can be calculated from the slope of the double reciprocal plot.

$$\text{slope} = 2.51 \times 10^6 \text{ (s M}^{-1}/\text{mM}^{-1})$$

$$\text{slope} = \frac{K_m(1 + I/K_i)}{V_m}$$

$$(2.51 \times 10^6 \text{ s M}^{-1}/\text{mM}^{-1}) = \frac{0.67 \text{ mM}(1 + I/K_i)}{5.0 \times 10^{-7} \text{ M s}^{-1}}$$

$$(1 + I/K_i) = \frac{(2.51 \times 10^6 \text{ s M}^{-1}/\text{mM}^{-1})(5.0 \times 10^{-7} \text{ M s}^{-1})}{(0.67 \text{ mM})}$$

$$(1 + I/K_i) = 1.87$$

$I/K_i = 0.87$ but [I] was 10µM so:

$$K_i = \left(\frac{10 \text{ µM}}{0.87}\right) = 11 \text{ µM}$$

Inhibitor B affects both the slope and the Y-intercept of the double reciprocal plot compared to the uninhibited enzyme, a pattern consistent with mixed or noncompetitive inhibition. The Michaelis-Menten expression, derived for noncompetitive inhibition, is

$$v_i = \frac{V_m(S)}{K_m(1 + I/K_{is}) + (S)(1 + I/K_{ii})}$$

The reciprocal form of the expression is

$$\frac{1}{v_i} = \frac{K_m}{V_m}(1 + I/K_{is})\frac{(1)}{(S)} + \frac{(1 + I/K_{ii})}{V_m}$$

K_i values can now be derived from the y-intercept (K_{ii}) and slope and (K_{is}) is as follows:

10. (continued)

  a. K_{ii}

$$Y\text{-intercept} = \frac{(1 + I/K_{ii})}{V_m}$$

$$0.42 \times 10^7 \text{ M}^{-1}\text{ s} = \frac{(1 + I/K_{ii})}{(5.0 \times 10^{-7} \text{ M s}^{-1})}$$

$$(1 + I/K_{ii}) = (0.42 \times 10^7 \text{ M}^{-1}\text{ s})(5.0 \times 10^{-7} \text{ M s}^{-1})$$

$$(1 + I/K_{ii}) = 2.1$$

$I/K_{ii} = 1.1$ but [I] was 10 μM, so

$$K_{ii} = 9.1 \text{ μM}$$

  b. K_{is}

$$\text{slope} = \frac{K_m (1 + I/K_{is})}{V_m}$$

$$\text{slope} = 2.8 \times 10^6 \text{ s M}^{-1}/\text{mM}^{-1}$$

$$2.8 \times 10^6 \text{ s M}^{-1}/\text{mM}^{-1} = \frac{0.67 \text{ mM}^{-1} (1 + I/K_{is})}{5.0 \times 10^{-7} \text{ M s}^{-1}}$$

$$(1 + I/K_{is}) = (2.8 \times 10^6 \text{ s M}^{-1}/\text{mM}^{-1})(5.0 \times 10^{-7} \text{ M s}^{-1})$$

$$1 + I/K_{is} = 2.1$$

$I/K_{is} = 1.1$ but [I] was 10 μM so

$$K_{is} = 9.1 \text{ μM}$$

In this example, K_{ii} and K_{is} are equal, but that is not usually the case.

12. The determination of the V_{max} in a velocity versus [S] plot (Figure 8.6) is often not trivial. Frequently the position of the asymptote is difficult to determine. The determination of a value for the K_m then depends on the selected value of V_{max}. The advantage of a double reciprocal (Figure 8.9) is that much of the decision making process in choosing V_{max} is simplified.

14.

Sample #	[2-Deoxyribose] mmolar	$\frac{1}{[S]}$	ADP formation, change in [ADP] in micro M/10 min.	$\frac{1}{V^0}$
1	1.00	1.00	0.115	8.70
2	1.00	1.00	0.124	8.06
3	1.60	0.625	0.147	6.80
4	2.56	0.391	0.191	5.24
5	5.12	0.195	0.255	3.92
6	6.88	0.145	0.274	3.65
7	15.0	0.066	0.294	3.40
8	20.0	0.050	0.300	3.33

From V_o versus [S] the V_{max} is 0.31 micromolar/10 min. and K_m is 1.85×10^{-3} M and from a Lineweaver-Burk plot the V_{max} is 0.33 micromolar/10 min. and the K_m is 1.8×10^{-3} M.

16. Organic chemists are more likely to deal with reactions in the gas phase or in solution in which the temperature is often elevated to increase the energy content of the reactants. Biochemists, on the other hand, typically visualize a reaction occurring at pH 7, 37°C and an aqueous environment on a protein (enzyme) surface involving intimate contact/proximity between the surface and the substrate(s).

18. Diffusion is random, which is difficult to reconcile with the apparent "perfection" of an enzyme's catalytic activity. Missing in our thoughts is the speed/distance relationships at which diffusion operates.

9   Mechanisms of Enzyme Catalysis

Chapter 9 Answers

2. Flexibility is a means that proteins can utilize to induce a sequential binding of
    substrates. If the binding site for the second substrate is formed by the enzyme binding
    of the first substrate then the structure of the protein dictates the order of binding of the
    substrates.

4. The amide N-H groups of ser 195 and gly 193 form hydrogen bonds to the carbonyl
    oxygen of the amide bond that is subject to hydrolysis. The hydrogen bonds facilitate the
    formation of and oxyanion intermediate but in the process must be able to cope with the
    geometry changes from an sp^2 to sp^3 in the carbonyl oxygen and carbon.

6. a. The proteolytic enzymes trypsin ,chymotrypsin, and elastase share the following
        mechanistic features.

    (a) *Each enzyme has the active site "catalytic triad" of aspartate, histidine, and
         serine.* In the currently proposed model for the serine protease mechanism, the
         aspartate is hydrogen-bonded to the imidazole group of histidine. The histidine
         accepts the proton from the serine hydroxyl group during nucleophilic addition
         of the serine —OH to the substrate peptidyl bond. The active-site aspartate
         and the oxyanion hole contribute to the electrostatic stabilization of the
         tetrahedral transition state intermediate. (See Warshel, A. et al., *Biochemistry*.
         28: 3629–3637, 1989.)

    (b) *Each forms an acyl-enzyme intermediate during peptide bond cleavage.* The
         nucleophilic OH group of serine adds to the carboxyl of the substrate peptide
         bond, and an acyl ester is formed between the serine OH and the N-terminal
         portion of the peptide substrate. The C-terminal portion of the substrate
         diffuses from the active site.

    (c) *The acyl-enzyme intermediate is deacylated by hydrolysis.* Addition of water
         hydrolyzes the acyl ester and restores the active site residues to their original
         chemical state. See figure 9.11 in text.

6. **(continued)**

    b. The specificity of each enzyme is dictated by the size and relative hydrophobicity of the substrate binding crevice that is adjacent to the catalytically active serine residue.

        Trypsin cleaves peptide bonds on the carboxyl side of internal lysine or arginine residues. The charged, polar side chains of those amino acids are bound in crevice that has an acidic residue, aspartate, located at its base. Electrostatic interaction occurs between the aspartate and the positively charged side chains of the Lys or Arg residues.

        Chymotrypsin preferentially cleaves peptide bonds on the carboxyl side of internal Phe, Tyr, or Trp residues. These amino acid side chains are hydrophobic aromatic (Phe and Trp) or polar but not charged (Tyr). The active site crevice that binds these residues is considerably more hydrophobic than is the binding pocket in trypsin.

        Elastase cleaves on the carboxyl side of the less bulky, uncharged residues, such as alanine or valine. The smaller size of the side chains is accommodated in a shallower, active site crevice.

8. Water structure and its thermodynamic contribution to biochemical reactions must be considered in understanding reaction mechanisms. Water molecules that surround hydrophobic residues or solvate ionic amino acid side chains in the enzyme active site are less randomly distributed than are water molecules in bulk solution (see chapter 2 in the text). Substrate binding to the enzyme is frequently accompanied by a conformational change of the protein concomitant with expulsion of water from the active site. The conformational change fosters hydrophobic and ionic interactions between the enzyme active site and the substrate and ultimately between enzyme and transition-state intermediate. The decreased organization of the water molecule expelled from the active site increases the entropy (ΔS) of the system and decreases ΔG. Moreover, the pK_a values of acidic or basic functional groups at the active site may differ significantly from those of the free amino acids because removal of water favors the uncharged forms.

10. Proteins are assembled from amino acids into the primary sequence based on information encoded in the gene sequence. The primary sequence dictates the formation of secondary structures, and these structures fold and interact to form the tertiary structure. Renaturation of denatured protein is also dictated by the primary structure of the protein.

    The trypsin family of enzymes and carboxypeptidase A are each synthesized as proenzymes that are proteolytically activated after synthesis. The proteolyzed, active enzymes have primary structures that differ from the gene product (proenzyme) and would not be expected to refolded into a conformation equivalent to that of the proenzymes. It is not surprising that the renatured enzyme is not catalytically active. In addition, zinc is a cofactor required for carboxypeptidase A activity and must be available during the renaturation process. Addition of Zn^{2+} to the renaturing proenzyme would likely result in the correct ligand attachment to the metal. The correct placement of the ligands may not occur in the refolding of denatured, proteolytically activated enzymes.

12. The enzyme active site stabilizes the transition-state intermediate formed during catalysis with a resulting decrease in the activation energy of the reaction. The structure of the transition-state intermediate is complementary to the structure of the active site and is thus bound much more tightly than are either substrates or products. Transition-state analogs, having structural features only slightly different from the transition-state intermediate, would be expected to bind very tightly to the active site and thus have a low K_i. The analog would compete with substrates or product for the active site and would exhibit competitive inhibition.

10 Regulation of Enzyme Activities

Chapter 10 Answers

2. Most students find allosteric models to be simpler. Covalent modification requires an enzyme to control an enzyme. (How do you control the first enzyme?) Allosterism requires only a binding site on an enzyme which interacts (via an equilibrium) with a particular small molecule.

4. The first enzyme of a pathway produces the first substance that is unique to only that pathway. To regulate the first enzyme makes sense because it involves the smallest commitment of materials and energy. Does it make sense for an automobile manufacturer to control its car production at the "addition of the hubcap" stage?

6. The methylene group prevents the elimination of a phosphate, which is required to convert the postulated intermediate into carbanoyl-aspartate. The suggested oxygen analog might eliminate phosphate (i.e., be a substrate for aspartate carbamoyl transferase).

8. Both processes utilize ATP and an enzyme-OH is modified. Different (complementary) portions of the ATP are utilized to produce a phosphorylated and adenylated enzyme. The enzyme-OH is provided by a Tyr, Ser, or Thr for phosphorylation and a Tyr for adenylation.

10. Phosphorylation of serine or threonine (and possibly tyrosine) on the target protein may be largely influenced by the amino acid sequence around these residues. These amino acid sequences may define a specific motif or recognition site for the protein kinase. Kemp (*TIBS*. 15:342–346 (1990)) and Kennelly and Krebs (*J. Biol. Chem.* 266:15555–58 (1991)) have reviewed possible recognition sequences (motifs) on the substrates for a number of protein kinases. The authors caution that not all serine (threonine) residues in a specific motif are necessarily phosphorylated. Serine (or threonine) residues amay not be phosphorylated because of other factors, such as topographical features on the target protein, that prevent binding of the target protein to the protein kinase active site.

12. Allosteric regulation of an enzyme having a binding site for a regulatory molecule and an active site on the same subunit is not uncommon. Regulatory molecules bind at sites separate from the active site and induce a conformational change that affects substrate binding to the active site on the same as well as adjacent subunits.

14. Thioredoxin alters the activity of target proteins by specific disulfide bonds to dithiols.

$$E(\text{—S—S—}) + Tr(SH)_2 \rightarrow E(SH)_2 + Tr(\text{—S—S—})$$

where $E(\text{—S—S—})$ and $Tr(\text{—S—S—})$ are the oxidized forms and $E(SH_2)$ and $T(SH)_2$ are reduced forms of the target protein or target enzyme and thioredoxin ,respectively. Oxidized thioredoxin is reduced in the plant cell by reduced ferredoxin.

$$2\ Fd_{red} + Tr(\text{—S—S—}) \rightarrow 2\ Fd_{ox} + Tr(SH)_2$$

It should be noted that two moles of ferredoxin are required per mole of oxidized thioredoxin reduced because ferredoxin is a one electron-equivalent redox reagent, whereas reduction of the disulfide requires two electrons.

11 Vitamins and Coenzymes

Chapter 11 Answers

2. Schiff Base formation between lysine and pyridoxal phosphate does not involve the alpha hydrogen. Upon decarboxylation the lone pair of electrons on the alpha carbon is resonance stabilized back into the pyridoxal phosphate. The lone electron pair on the alpha carbon picks up a hydrogen ion and following Schiff base hydrolysis cadaverine is formed without the loss of the original alpha hydrogen. Racemases and transaminases involve a mechanism in which the alpha hydrogen on the amino acid is released.

4. Lipoic acid and biotin are each covalently bound to their respective enzymes through amine bonds involving the carboxyl group of the cofactor side chain and the e-amino group of lysine in the protein. The combined length of the lysyl side chain and the acyl group on the cofactor allows free rotation of the dithiol group (lipoic acid) or the imidazolone group (biotin) at the end of a tether that may extend to approximately 14 to 16 Å long. The chemically reactive portion of the cofactor may diffuse between several catalytic centers present either on different subunits of the enzyme or different enzymes of a multienzyme complex but the cofactor is not physically released from the complex. Moreover, the effective concentration of the cofactor at the active site is greater than if the equivalent amount of cofactor were free in solution, thus increasing the probability of a chemically reaction.

6. Pyridine nucleotide-dependent dehydrogenases typically have almost absolute specificity for either NAD^+ or $NADP^+$ as a cofactor. The specificity of the dehydrogenases provides a way to differentiate anabolic and catabolic processes. For example, the mitochondrial β-oxidation of fatty acids requires NAD^+ as an oxidant of the β-hydroxyacyl group to the β-ketoacyl group. In fatty acids biosynthesis, NADPH is used to reduce the β-ketoacyl group to a β-hydroxyacyl group and to reduce the β-enoyl group.

   The ratios of $NADH/NAD^+$ and $NADPH/NADP^+$ may contribute to differential regulation of metabolic pathways.

8. a.

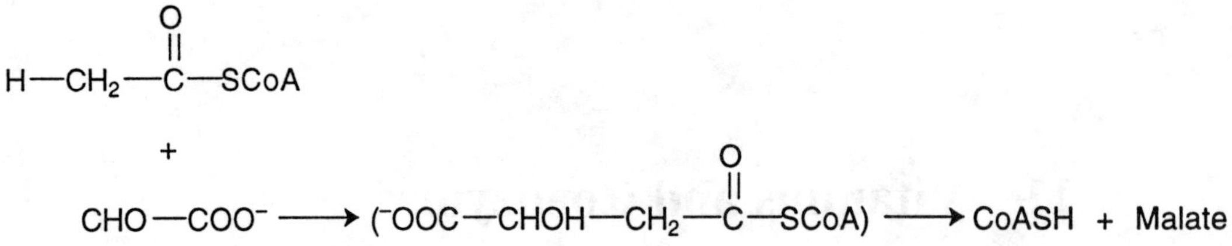

$$H-CH_2-\overset{\overset{\displaystyle O}{\|}}{C}-SCoA$$

$$+$$

$$CHO-COO^- \longrightarrow (^-OOC-CHOH-CH_2-\overset{\overset{\displaystyle O}{\|}}{C}-SCoA) \longrightarrow CoASH + Malate$$

b. Malate synthase catalyzes the Claisen condensation of acetyl-CoA and glyoxalate. In that reaction, an α-proton is abstracted and the carbanion thus formed adds to the aldehyde carbonyl group of glyoxalate. The pK of the α proton is decreased (the proton is more acidic) in the acetyl-CoA thioester compared to free acetate. α-carbanions of thioesters are more stable than if the carboxyl group were either in an oxygen ester or were unesterfied. Abstraction of an α proton from acetate, forming the dianion, is much less likely to occur. (See Walsh, C. *Enzymatic Reaction Mechanisms*. San Francisco: W.H. Freeman, 1977, 759–62, *and* Higgins, M.J.P., J.A. Kornblatt, and H. Rudney in *The Enzymes*, Vol. VII, ed. P. Boyer. New York: Academic Press, 1972, 412–22.)

c. Coenzymes A is a good leaving group and hydrolysis of the thioester is thermodynamically favorable. Hydrolysis of the thioester contributes approximately −10 kcal/mole to the overall reaction. Thus, the products malate and CoASH are more stable when compared with the malylthioester.

10. a. Acetyl-CoA carboxylase

   (1)   Carboxylation of biotin

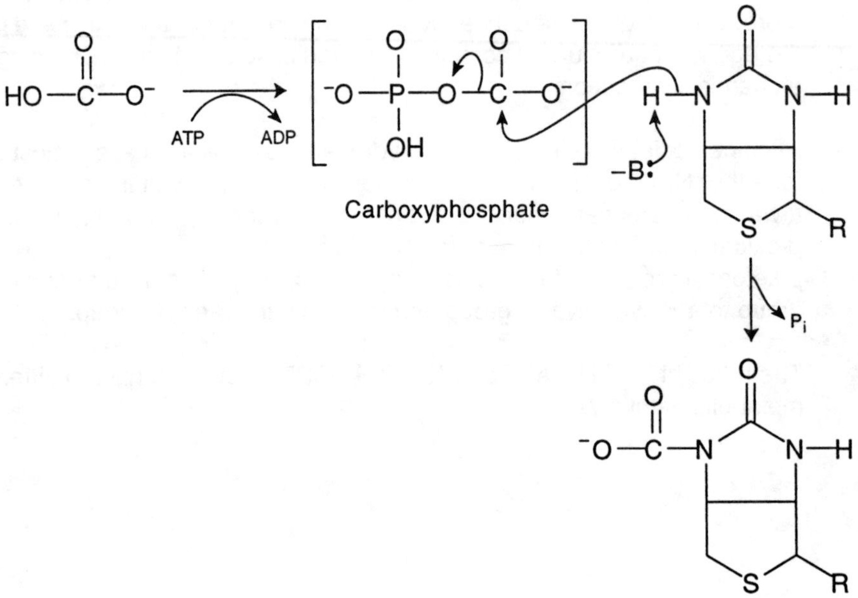

10. a. (continued)

(2) Proton abstraction from methyl of acetyl-CoA, addition of carbanion to carboxyl group of carboxybiotinyl enzyme.

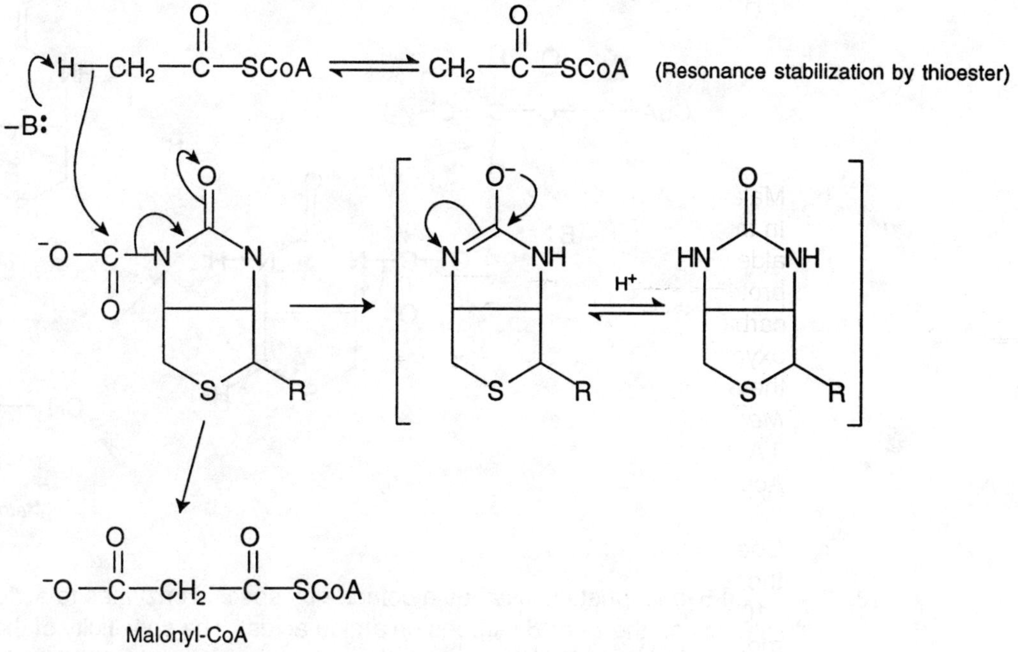

Malonyl-CoA

b. Transcarboxylase

(1) Transfer of β-carboxyl group from oxaloacetate to biotin

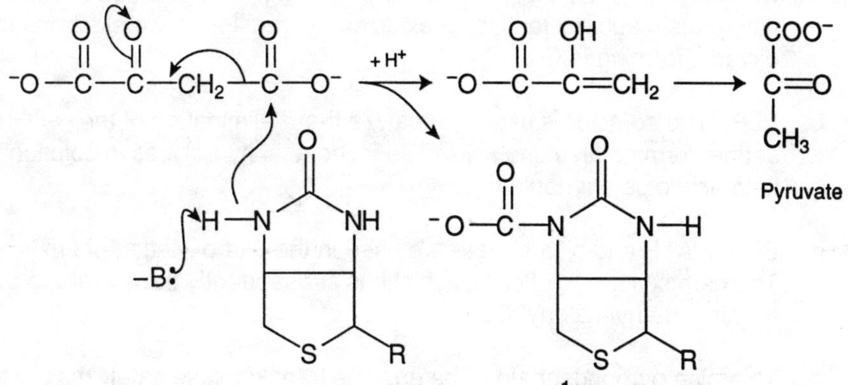

Pyruvate

N^1-carboxybiotin

10. b. (continued)
   (2)   Transfer of carboxyl group from carboxy biotin to propionyl-CoA

Methyl malonyl-CoA

12. Pyridoxal-5'-phosphate is used as a cofactor by specific enzymes to activate substituents on the α- or β-carbons on amion acids. The specificity of the catalyst, the enzyme, dictates which of the activated bonds will specifically be altered. Thus, amino transferases utilize PLP to remove the α-amino group without α-decarboxylation as an interfering reaction.

14. a. Flavin. The enzyme catalyzes the oxidation of the α carbon, forming an α-imino acid with transfer of two electrons to the flavin ring. The reduced flavin on the flavoprotein subsequently is reoxidized by transfer of two electrons to molecular oxygen, forming H_2O_2.

   b. PLP. The cofactor is used to catalyze the β-elimination of the —OH group from serine, forming an imino acid. The imino acid hydrolyzes in solution to yield the α-keto acid plus ammonium ion.

   c. Biotin. ATP and bicarbonate are used in the carboxylation of enzyme-bound biotin. The carboxyl group of carboxybiotin is subsequently transferred to propionyl-CoA, forming methylmalonyl-CoA.

   d. Thiamine pyrophosphate. The enzyme transketolase catalyzes the transfer of C1-C2 as a unit from fructose-6-phosphate to thiamine pyrophosphate, with the release of the 4-carbon sugar, erythrose-4-phosphate. The 2-carbon unit bound to thiamine pyrophosphate is subsequently transferred to an acceptor, glyceraldehyde-3-phosphate. The product is the 5-carbon sugar, xylulose-5-phosphate.

16. NAD$^+$ is cleaved at the N-glycosyl bond between nicotinamide and the ribose ring. The cleavage of this bond releases approximately 8 kcal/mole and is thermodynamically favorable. Adenosine diphosphate ribose is transferred to the acceptor, and nicotinamide and a proton are released. ADP-ribosylation or poly ADP-ribosylation is a posttranscriptional modification of the acceptor protein in which NAD$^+$ is consumed as a substrate in the reaction, rather than functioning as a cofactor (1). ADP- or polyADP-ribosylation of several proteins has been documented, including polyADP-ribosylation of nuclear proteins and ADP-ribosylation of elongation factor-2, catalyzed by the toxin of *Corynebacterium diphtheriae* (2,3).

18. The hydroxyl hydrogen on carbon 3 is the most acidic because the resulting anion is resonance stabilized to a greater degree than the hydroxyl hydrogen on carbon 2.

20. Having required coenzymes covalently bound to the enzyme facilitates the catalysis by allowing the coenzyme to be continually available. Both biotin and lipoic acid have "tails" consisting of a four carbon chain terminated in a carboxylic acid. The amide linkages observed create a ten atom long "arm" which allows the functional part of the coenzyme to "visit" a substantial portion of the enzyme's surface. This latter feature is vital in the mechanism of the enzymes that use these coenzymes. What other amino acid residues R-groups could be used for the covalent attachment of these "tails" ending with a carboxylic acid? Both ester linkages (to seryl or threonyl residues) or anhydrides (oto glutamyl or aspartyl residues) can be envisioned but these linkages are too energetic and the coenzyme could be lost from the enzyme by an undesired non-enzymatic side reaction.

22. (a) redox agents: FAD, FMN, NAD$^+$, NADP$^+$, and lipoyl; (b) acyl carriers: CoASH, lipoyl, and thiamine pyrophosphate; (c) both acyl carriers and redox agents: lipoyl.

12   Metabolic Strategies

Chapter 12 Answers

2. Futile cycles could (typically) occur when enzymes for: $ATP + ROH \rightarrow ADP + R-P_i$ and $R-Pi + H_2O \rightarrow ROH + P_i$ are combined in the same cell. As long as the two enzymes are *not active* at the same time there is no futile cycle.

4. The compound that accumulates depends upon the equilibrium position of the previous reactions in the pathway.

6. The transition from left to right involves dehydration reactions, while from right to left involves hydrolysis reactions. The items on the left can be thought of as monomers while only the first three items on the right are polymers. A typical lipid contains fatty acids but is not considered to be a polymer.

8. Catabolism involves pathways composed of enzymes and chemical intermediates that are primarily involved in the breakdown of large molecules into small molecules, often by oxidation processes. Anabolism is the collection of the enzymes and chemical intermediates involved in the biological synthesis of larger molecules from smaller molecules. Anabolism often involves reduction processes. Collectively anabolism and catabolism are called metabolism.

10. The advantage of subcellular compartments is that a specific pathway can be isolated from comparable pathways that might use similar or identical chemical intermediates. This simplifies regulation of the various processes.

12. a. The concentration of most metabolites measured under steady-state conditions in the cell usually does not exceed the K_m value. For an enzyme whose reaction can be described by simple Michaelis-Menten kinetics, the observed velocity, v, is 0.5 V_{max} if substrate concentration equals the K_{max} value. The velocity of most enzymes is likely significantly less than V_{max} *in vivo*.

b. End-product inhibition usually occurs at the committed step in a metabolic pathway or at a branch point in the pathway. Regulation of the enzyme(s) at these steps in the metabolic pathway prevents the accumulation of intermediates in the pathway when the cell has limited demand for the end product. Regulation by individual metabolites at a branch point of the pathway inhibits their own production without simultaneously inhibiting the production of product from the other branch. Each product may cumulatively inhibit an enzyme catalyzing a precursor common to each end product.

12. (continued)

    c. Catabolic pathways tend to be convergent rather than divergent. Metabolic convergence of precursors into common intermediates of a metabolic pathway provides an efficient route for the metabolism of a variety of metabolites by a limited number of enzymes. For example, the catabolism of glucose, fructose, and glycerol to the common intermediate glyceraldehyde-3-phosphate in the liver illustrates the convergence of the glycolytic pathway. Anabolic pathways tend to be divergent, with the synthesis of several end products from a common precursor.

    d. Enzymes that are regulated in metabolic pathways most frequently exhibit cooperative kinetic responses rather than hyperbolic response with respect to substrate concentration and are frequently responsive to allosteric regulation by products, energy charge, or concentration ratio of NAD(P)H/NAD(P)*. The rate of enzymatic activity over a narrow range of substrate concentration can be changed dramatically by allosteric activators or inhibitors. Such dramatic changes in velocity in response to small changes in substrate concentration are not observed with enzymes exhibiting Michaelis-Menten kinetics.

    e. Energy charge is a means of expressing the fraction of adenylate nucleotide that are high free energy compounds: ATP and ADP. The expression

$$\text{E.C.} = \frac{[\text{ATP}] + [\text{ADP}]/2}{[\text{ATP}] + [\text{ADP}] + [\text{AMP}]}$$

where E.C. represents the energy charge varying theoretically between 1 (all ATP) to 0 (all AMP). The energy charge is metabolic signal for both anabolic (biosynthesis) and catabolic (degrative) metabolic pathways. Low energy charge signals the cell that a need for ATP formation exists, and pathways (glycolysis and Krebs cycle) leading to ATP formation are activated. Anabolic pathways that demand high ATP concentrations are inhibited at low energy charge. In the latter case, the ATP required to drive biosynthesis is in low supply. Conversely, increased energy charge inhibits pathways leading to ATP formation and activates anabolic pathways.

    f. Formation of multienzyme complexes is a strategy frequently used for efficient catalysis and control of metabolic pathways. Substrates diffuse shorter distances between active sites in multienzyme complexes than if the enzymes were not organized. Frequently, intermediates covalently bound to cofactors (e.g., lipamide, biocytin) are moved among active sites within the complex, effectively trapping intermediates in the complex and increasing the concentration of substrates at the enzyme active site.

    g. Separation of catabolic and anabolic pathways diminishes the likelihood of futile cycling of metabolites. Enzymes catalyzing the β-oxidation of fatty acids are located in the mitochondrial matrix, whereas enzymes catalyzing the synthesis of palmitate are located in the cytosol.

14. The "committed step" in a reaction sequences steers the metabolite to a sequence of reactions whose intermediates usually have no other function in the cell. Control of the committed step prevents wasteful accumulation of these single-purpose intermediates and obviates the necessity of controlling each enzyme in a pathway. Enzymes that catalyze reactions just past a branch point in a pathway are likely inhibited by their respective products or the end product of the branch.

13 Structures of Sugars and Energy-Storage Polysaccharides

Chapter 13 Answers

2. Mannose is the 2-epimer and galactose the 4-epimer of glucose, which means that mannose and galactose each have only one "bulky" group in an axial position.

4. The sugar is alpha. Only the OH on carbon-2 is in the axial position for the following sugar:

6. Polysaccharides exist, in some instances, as highly branched structures. The number of different structural arrangements possible in polysaccharides is staggering. When one considers higher order polymers of the various sugars, the diversity of structures may exceed the possible arrangement of linear amino acids in polypeptides. However, the carbohydrates may be unable to provide the diversity of chemical reactivity or binding characteristics that can be provided by the amino acids. Amino acids provide diverse functional side chains (strongly basic, acidic, and hydrophobic entities) that carbohydrates lack. In addition, one would have to consider information transfer to allow specific synthesis of catalytic carbohydrates. Amino acids are linked into linear polypeptide chains under the direction of a linear sequence of nucleotides. construction of branched carbohydrates from a linear template would be difficult.

8. Solutions of pure α or β anomer are thermodynamically less stable than the final mixture. The anomers will mutarotate until a lower energy steady-state mixture is achieved. Each anomeric form undergoes reversible opening and reclosing of the pyranose ring, with the equilibrium position strongly favoring ring closure. In the ring-open configuration, the aldehyde group may rotate so that either side (face) of the carbonyl is available for readdition of the —OH group from C-5, thus generating the mixture of anomers.

10.

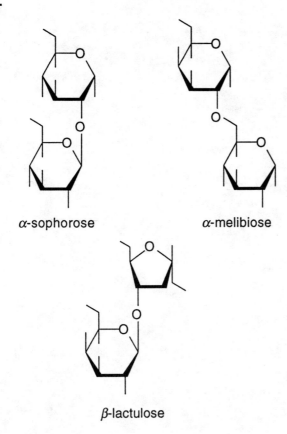

α-sophorose   α-melibiose

β-lactulose

12. As the uro-root indicates, glucuronic acid was first found as part of glycolysis in urine.

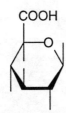

β-*D*-Glucuronic Acid

14.

major part of inulin

14 Glycolysis, Gluconeogenesis, and the Pentose Phosphate Pathway

Chapter 14 Answers

2. There is a substantial energy barrier to overcome in the conversion of pyruvate to phosphoenolpyruvate. Some of the energy provided by ATP is preserved in the carbon-carbon bond formed during the conversion of pyruvate into oxaloacetate. This energy preserved in the carbon-carbon bond is released in the loss of the CO_2 in the next step (phosphoenolpyruvate carboxykinase) and together with an additional ATP there is sufficient energy available for the formation of phosphoenolpyruvate. The CO_2 added by pyruvate carboxylase is the same CO_2 as lost in the phosphoenolpyruvate carboxykinase reaction.

4. The enzymatic reaction from phosphoenol pyruvate to pyruvate is irreversible. The enzyme's name describes the reverse reaction.

6. One possibility is to use the pentose phosphate pathway to glyceraldehyde-3-phosphate and then send the latter compound through glycolysis. Any fructose-6-phosphate produced in the pentose phosphate pathway could be routed through glucose-6-phosphate to the pentose phosphate pathway.

8. a. Bacteria are adept at fermenting numerous sugars in addition to glucose. The sugars are used as carbon source for metabolic energy transduction and for biosynthesis.

    b. Glycolytic conversion of glucose to glyceraldehyde-3-phosphate ($Ga3P_i$) occurs without net oxidation or reduction. $Ga3P_i$ is oxidized to a thioester that remains covalently bound to the enzyme ($Ga3P$ dehydrogenase) with the concomitant reduction of the electron transfer coenyzme, NAD^+, to NADH. The 3-phosphoglyceryl group is released from the enzyme by phosphorolysis and is subsequently metabolized to pyruvate. Pyruvate, the α-keto acid., is reduced by the NADH-dependent lactate dehydrogenase in order to resupply NAD^+ to Ga3PDH. Hence, there is no net oxidation, but the product (lactate) is of lower free energy than the substrate (glucose). The energy difference between product and substrate is used to drive the phosphorylation of ADP.

8. **(continued)**

    c. No CO_2 is released upon conversion of glucose to lactate. However, the culture could also be decarboxylating pyruvate to acetaldehyde and CO_2. The acetaldehyde may subsequently be reduced by the NADH-dependent alcohol dehydrogenase to ethanol using the NADH from the glyceraldehyde-3-phosphate dehydrogenase reaction. As noted in part (b), pyruvate is the terminal electron acceptor in glycolysis. During fermentation, acetaldehyde is the terminal electron acceptor. In either case, NAD^+ is resupplied to the Ga3PDH to prevent inhibition of glycolysis at the Ga3P$_i$ dehydrogenase step.

10. a. Triose phosphate isomerase deficiency would inhibit conversion of DHAP to Ga3P and would cause accumulation of DHAP, preventing half of the glucose molecule (C1-C3) from being metabolized through the remainder of the glycolytic pathway. There would be a recovery of only 2 of the possible 4 moles of ATP from glucose, resulting in no net formation of ATP. In addition, DHAP, a product of the aldolase reaction, would likely reverse the aldolase (reaction) and eventually inhibit glycolysis. Either result would be lethal to a cell whose only energy source was glycolysis.

    b. The small amount of TPI activity would likely allow glycolysis to proceed slowly, but low energy (ATP) level will limit the growth rate under anaerobiosis. However, the yield of ATP is significantly greater when the pyruvate, formed during glycolysis, is oxidized to CO_2 and H_2O. Hence, the growth rate of the mutant should be correspondingly greater under aerobic growth conditions but not as great as the wild type.

    c. Cells expressing DHAP phosphatase would likely not grow anaerobically if glycolysis of glucose to lactate were the only pathway for ATP formation. The combined activities of TPI and DHAP phosphate would be predicted to deplete the pool of triosephosphate and the yield of ATP per glucose would likely be less than 1. The cells might grow aerobically, depending on the competition among Ga3PDH, TPI, and DHAP phosphatase for the triosephosphate pool.

    Glycerol can be phosphorylated to α-glycerolphosphate and oxidized to DHAP. The organism expressing DHAP phosphate would likely not grow anaerobically on glycerol, but might grow aerobically for the reasons described above.

12. a. Consider the reactions:

    $$H_2O + \text{Sucrose} \rightarrow \text{fructose} + \text{glucose} \qquad \Delta G^{\circ\prime} = -7 \text{ kcal/mole}$$

    $$\text{Glucose} + P_i \rightarrow \text{lucose-1-}P_i + H_2O \qquad \Delta G^{\circ\prime} = +5 \text{ kcal/mole}$$

    $$\text{Sucrose} + P_i \rightarrow \text{Glucose-1-}P_i + \text{fructose} \qquad \Delta G^{\circ\prime} = -2 \text{ kcal/mole}$$

    $\Delta G^{\circ\prime} = -2.3RT \log K'_{eq}$ where $2.3\,RT$ is 1.36 kcal/mole at 25°C.

    $$-2 \text{ kcal/mole} = -1.36 \text{ kcal/mole} \log K'_{eq}$$

    $$\text{Log } K'_{eq} = 1.47$$

    $$K'_{eq} = 30$$

12. **(continued)**

   b. Hexoses brought into the glycolytic pathway must be phosphorylated to provide the appropriate substrate for the glycolytic enzymes and to trap the sugar within the cell. Phosphorylation at the expense of ATP or group translocation at the expense of PEP are common methods to activate the sugar molecules. Sucrose phosphorylation uses the exergonic lysis of glycosidic bond between the hemiacetal OH group of glucose and the hemiketal OH group of fructose to drive the endergonic phosphorylation of the hemiacetal C-1 OH group of glucose. Transfer the phosphate to C-6, catalyzed by phosphoglucomutase, provides substrate for entry into the glycolytic pathway without addition of ATP. The ATP yield will therefore be 3 rather than 2 moles of ATP per mole of glucose derived from sucrose. The fructose can be phosphorylated by ATP and used in the glycolytic pathway.

14. PEP formed by the PEP carboxykinase reaction is hydrated by enolase to form 2-phosphoglycerate (2-PGA) whose C-2 is a chiral center. Conversion of 2-PGA to 3-phosphoglycerate by phosphoglyceromutase, phosphorylation by phosphoglycerokinase, and reduction to Ga3P$_i$ by Ga3PDH retains the chirality at C-2 but introduces not additional chrial centers. Aldolase-catalyzed condensation of DHAP, formed from the Ga3P$_i$, with Ga3P$_i$ introduces chiral centers at C-3 and C-4 of the fructose-1,6-bisphosphate. The chiral center at C-5 arises from C-2 of Ga3P$_i$. The remaining chiral center at C-2 is introduced upon conversion of the Fru-6-P$_i$ to Glc-6-P$_i$ catalyzed by phosphoglucoisomerase.

16.

6-phosphogluconate

Ribulose-5-P$_i$

15 The Tricarboxylic Acid Cycle

Chapter 15 Answers

2. If the oxaloacetate formed leaves the cycle to produce pyruvate, then acetyl-CoA, there will be a loss of carbon-4 then carbon-1 of the original aspartate. If the oxaloacetate condenses with an acetyl-CoA then there will be a loss of carbon-1 of aspartate at the isocitrate dehydrogenase stage.

4. The oxidation occurs first, creating a β-keto carboxylate which readily loses CO_2 via a resonance stabilized carbonion.

6. Citrate is a tertiary alcohol, which does not oxidize readily. Isocitrate, a secondary alcohol, readily oxidizes to a ketone (an intermediate in fig. 15.9).

8. The following sequence will do the task: α-ketoglutarate, the tricarboxylic acid cycle to oxaloacetate, to phosphoenolpyruvate, to pyruvate, to acetyl-CoA, into the tricarboxylic acid cycle.

10. Both of the oxidative-decarboxylation steps in the tricaboxylic acid cycle are by-passed by the glyoxylate cycle.

12. a. Increased concentrations of acetyl-CoA slows the activity of pyruvate dehydrogenase.

    b. The removal of succinyl-CoA for heme synthesis removes any control it has over lowering the activity of alpha ketoglutarate dehydrogenase and citrate synthetase.

14. a. False. Pyruvate dehydrogenase catalyzes reduction of the lipoamide disulfide concomitantly with oxidation and transfer of the hydroxyethyl group from thiamine pyrophosphate. The hydroxyethyl group is oxidized to form a thioester with one sulfhydryl group of lipoamide. The acetyl group is subsequently transferred to CoASH, forming the thioester of adduct, acetyl-CoA. During the oxidation of the α-keto carbon, the disulfide (oxidized form) of lipoamide is reduced to the dithiol form. Dihydrolipoamide dehydrogenase, a flavoprotein, catalyzes oxidation of dihydrolipoamide and reduction of NAD^+ to NADH.

    b. False. Hydrolysis of acetyl-CoA thioester should yield as much free energy as succinyl-CoA hydrolysis. Succinate thiokinase catalyzes the substrate level phosphorylation of GDP at the expense of succinyl-CoA hydrolysis. However, in the TCA cycle, there is no enzymatic pathway to couple the hydrolysis of the acetyl-CoA to the activation of orthophosphate and subsequent transfer of the activated phosphate to ADP. Succinate thiokinase provides such as enzymatic pathway to couple the energy of succinyl-CoA hydrolysis to GDP.

14. (continued)

   c. False. The TCA cycle is a versatile pathway for the oxidation of various substrates, including carbohydrates, fatty acids, and lipids. Thus, the methyl group of acetyl-CoA could be derived from pyruvate, from β-oxidation of long chain fatty acids, or from amino acid catabolism.

   d. False. The CO_2 molecule released by oxidative decarboxylation of isocitrate (ICDH) is derived from the carboxyl group of oxaloacetate with which the acetyl-CoA was condensed. However, the CO_2 released form α-keto glutarate does depend on discrimination between the ends of the citrate molecule. If the aconitase reaction were random, half the CO_2 would arise from an oxaloacetate carboxyl group and half from the acetate carboxyl group. Such is not the case because aconitase discriminates between the two arms of citrate.

   e. False. Malate can easily be dehydrated to fumarate by reversal of the fumarase reaction.

16. a. Without malate synthase, the yeast would be unable to grow on 2-carbon precursors as sole carbon source because TCA cycle intermediates could not be synthesized. The glyoxalate bypass provides a pathway for the net synthesis of malate from acetyl-CoA. Entry of malate, a dicarboxylate intermediate in the TCA cycle, would provide increased concentrations of each of the TCA cycle intermediates with subsequent increase in oxidation of acetyl groups. ATP formation by oxidative phosphorylation could then occur. In addition, the glyoxalate bypass provides a carbon source of synthesis of carbohydrate and precursors of other cellular constituents. The absence of malate synthase would adversely affect those reactions.

   b. Pyruvate carboxylase catalyzes the carboxylation of pyruvate to form oxaolacetate. Oxaloacetate thus formed enters the TCA cycle to replenish cycle intermediates. Pyruvate carboxylase is normally activated by acetyl-CoA; thus the need for oxaloacetate is tied to an increased supply of acetyl-CoA. Were the pyruvate carboxylase in the mutant less responsive to acetyl-CoA, it is possible that the TCA cycle activity would markedly diminish if the cycle intermediates were being used in biosynthetic pathways. In addition, the biosynthetic pathways (lipids, amino acids, and carbohydrates) dependent on TCA cycle intermediates would also be inhibited due to lack of metabolites.

   c. If the PDH were inhibited more strongly than usual by acetyl-CoA, one might suspect that acetyl-CoA concentration in the mitochondrial matrix would markedly decrease, in turn limiting activity of citrate synthase and diminishing TCA cycle activity. Hence the organism may become growth-limited because of lowered energy production and because of diminished concentrations of biosynthetic precursors supplied by the TCA cycle.

18. a. Fumarate + NADH + H$^+$ → succinate + NAD$^+$.

b. Fumarate is formed from the oxidation of succinate, a TCA cycle intermediate formed from the decarboxylation of α-ketoglutarate. α-Ketoglutarate dehydrogenase is an integral enzyme in the TCA cycle and is active aerobic metabolism of acetyl-CoA. However, under anaerobic conditions in *E. coli*, there is little, if any, α-ketoglutarate dehydrogenase activity. Succinate and therefore fumarate, are not produced by this route. Some reactions of the TCA cycle are reversible, however. For example, the reduction of oxaloacetate, formed by carboxylation of PEP pyruvate, yields *L*-malate, which may be dehydrated to fumarate. These reversible reactions form fumarate even if succinate dehydrogenase or α-ketoglutarate dehydrogenase reactions are blocked. (See Spiro, S., and J.R. Guest, *TIBS*. 16:310–314 (191).) Consider the reactions

Phosphoenolpyruvate + CO$_2$ → oxaloacetate + P$_i$
(PEP carboxylase)

Oxaloacetate + NADH + H$^+$ → *L*-malate + NAD$^+$
(malate dehydrogenase)

L-Malate → fumarase + HOH
(fumarase)

Fumarate + NADH + H$^+$ → succinate + NAD$^+$
(fumarate reductase)

c. In the reactions shown in part (6b), four reducing equivalents (two hydride groups) are transferred to carbon acceptors. Reduction of oxaloacetate to malate by malate dehydrogenase (MDH) and reduction of fumarate to succinate by fumarate reductase each requires hydride (or equivalent) transfer from NADH to the organic substrate. In the reduction of pyruvate to lactate via LDH only one hydride is used. Thus, two equivalents of NAD$^+$ are resupplied to glycolysis by the activities of MDH and fumarate reductase, whereas only one equivalent of NAD$^+$ is regenerated by LDH. Fumarate is one of the terminal electron acceptors used during anaerobic respiration in *E. coli*. However, each mole of PEP carboxylated is at the expense of 1 mole equivalent of ATP that could have been formed as a product of pyruvate kinase.

20. The committed step in a metabolic pathway is usually under metabolic control. Inhibition of the committed step in a metabolic sequence or pathway prevents the accumulation of unneeded intermediates and effectively precludes activity of the enzymes using those intermediates as substrates. The decarboxylation of pyruvate and the oxidative transfer of the hydroxyethyl group by pyruvate dehydrogenase constitutes the committed step in the pyruvate dehydrogenase catalytic sequence and is a logical control point.

16 Electron Transport, Proton Translocation, and Oxidative Phosphorylation

Chapter 16 Answers

2. The electrons from ascorbate enter the electron transport scheme before Complex IV. There is one coupling site between Complex IV and O_2 and this site has a P/O ratio of 1, so one ATP is produced per ascorbate. The production of ATP via the oxidation of ascorbic acid is not accepted as the normal role of vitamin C. Ascorbic acid is thought to participate in the formation of protein bound hydroxyproline from protein bound proline and possibly similar reactions with other amino acid residues.

4. a. Heme of the mitochondrial b-type cytochrome interacts hydrophobicaly with adjacent hydrophobic residues from the membrane-spanning α helices. The heme iron is fully coordinated through two imidazole groups from histidines in the protein. Heme in the o-type cytochromes is covalently bound to the protein through thioesthers formed by the addition of cysteinyl sulfhydryl groups to the vinyl substitutions on the heme ring. The iron is also fully liganded to a nitrogen from the imidazole group of hisitidine and a sulfur from the thioether linkage of methionine providing the fifth and sixth ligands. Heme a differs from the protoheme (heme) of the b- and o-type cytochromes by substitution of a formyl group at ring position 8 and a 17-carbon isoprenoid chain at position 2. The hydrophobic isoprenoid chain provides added hydrophobic interaction between the heme a and the protein. Iron of heme a is fully coordinated, but heme a, has an open ligand position available for binding of oxygen.

    b. Carbon monoxide binds to the reduced heme a/a_3 and cyanide binds to the oxidized heme a/a_3 presumably at the oxygen-binding site; they inhibit transfer of electrons from reduced cytochrome c to O_3. Electron transport is effectively blocked and ADP phosphorylation ceases. Neither CO nor CN^- at low concentration interact with the cytochrome b or c heme iron because there is not open ligand position to the iron available in these heme proteins.

6. a. In biological systems, the iron-sulfur centers are obligatorily single electron donors/acceptors, regardless of the number of Fe atoms in the center or their initial oxidation state. For example, the 4 (Fe-S) cluster iron-sulfur proteins accept electrons from flavoproteins or quinones in only 1-electron transfer cycles.

43

6.   **(continued)**

b.   The 4Fe-4S cluster is found in iron-sulfur proteins that transfer electrons at low and at high potential. The reduction potential is a measure of the ease of addition of an electrons to the couple, compared to the standard hydrogen electrode. Thus the protein component of the iron-sulfur protein affects the reduction potential. For example, an electron-withdrawing environment at the Fe-S cluster in principle should yield a more positive reduction potential than if the Fe-S cluster were in an electron-donating environment. Spatial constraints and physical interaction between the protein and the iron-sulfur cluster may also affect the redox potential. Moreover, the initial redox state of the cluster affects reduction potential, for example, the oxidized form of high potential iron-sulfur protein cluster is ($3\ Fe^{3+}$–$1\ Fe^{2+}$) whereas the oxidized form of the low potential iron-sulfur cluster is ($2\ Fe^{2+}$–$2\ Fe^{3+}$).

8.   The standard reduction potential $E^{\circ\prime}$ changes by the ratio $\Delta E^{\circ\prime}/\Delta pH = -(60\ mV)(n_{H+}/n)$ where n_{H+} is the proton equivalents taken up per equivalent (n) of electrons transferred in the reaction.

$$E^{\circ} = E^{\circ\prime}\ (pH\ 7) - E^{\circ}\ (pH)$$

The standard reduction potential of cytochrome c is unaltered at either pH 6 or pH 8 because not protons are taken up in the reduction reaction. The iron in the heme s reduced from ferric to ferrous state.

Reduction of ubiquinone results in the addition of two equivalents of protons per 2 electron equivalents in reducing the quinone to the dihydroquinone. Thus at pH 6, ΔpH is $(7-6) = +1$, so

$$E^{\circ\prime} - E^{\circ}\ (pH\ 6)\ \ =\ \ -60\ mV/1$$

$$E^{\circ\prime}\ (pH\ 7) - E^{\circ}\ (pH\ 6)\ \ =\ \ -60\ mV$$

$$E^{\circ}\ (pH\ 6) = E^{\circ\prime} + 60\ mV = (110\ mV + 60\ mV) = +170\ mV$$

$$E^{\circ}\ (pH\ 6)\ \text{is}\ +170\ mV$$

At pH 8,

$$[E^{\circ\prime} - E^{\circ}\ (pH\ 8)]\ \ =\ \ -60\ mV/-1$$

$$E^{\circ\prime} - E^{\circ}\ (pH\ 8)\ \ =\ \ +60\ mV$$

$$E^{\circ}\ (pH\ 8)\ \ =\ \ E^{\circ\prime} - 60\ mV$$

$$=\ \ (110\ mV - 60\ mV)$$

$$E^{\circ}\ (pH\ 8)\ \ =\ \ 50\ mV$$

Consider the reaction $UQ + 2e^{-} + 2\ H^{+} \rightarrow UQH_2$. Decreasing pH (increasing H^{+} concentrations) favors the formation of UQH_2 by mass action. Hence the quinone should be more easily reduced to UQH_2 at pH 6 and the value of E° calculated is consistent with the prediction. Decreasing the proton concentration favors the oxidized form of the quinone again by mass action and the quinone should be more difficult to reduce.

10. Examine each step in the metabolism described in the problem and determine whether reduced cofactor (NADH or FADH) or high-energy compound (ATP or equivalent) is produced. Using the (P/O) ratios given in the problem, sum the high energy phosphates formed divided by the gram-atoms of oxygen reduced. The P/O ratio is sometimes expressed as P/2e^- or moles of high-energy phosphate formed per two electron equivalents transferred to O_2.

   a. *Oxidation of lactate*
- Lactate dehydrogenase: Lactate + NAD$^+$ → pyruvate + NADH
  (NADH is extrametochondrial and is oxidized by the α-glycerolphosphate shuttle with P/O = 1.5/1).

- Pyruvate dehydrogenase: Pyruvate + CoASH + NAD$^+$ → acetyl-CoA + NADH + CO$_2$
  (P/O = 2.5/1).

Oxidation of acetyl-CoA via TCA cycle
- Isocitrate dehydrogenase (NADH, P/O = 2.5/1)
- α-Ketoglutarate dehydrogenase (NADH, P/O = 2.5/1)
- Succinate thiokinase (GTP but no oxygen reduced, P/O = 1/0)
- Succinate dehydrogenase (FADH, P/O = 1.5/1)
- Malate dehydrogenase (NADH, P/O = 2.5/1)

In the complete oxidation of 1 moles of lactate, there are 14 moles of ATP (or equivalent) produced and 6 gram-atoms of oxygen reduced. The theoretical P/O ratio is 2.33.

   b. 2,4-Dinitrophenol is an uncoupler of oxidative phosphorylation. The weakly acidic uncoupler dissipates the mitochondrial H$^+$ gradient and thus prevents ADP phosphorylation via oxidative phosphorylation. The metabolic energy is released as heat. Electron transfer to O_2 will proceed more rapidly in the uncoupled mitochondria than in tightly coupled mitochondria but without ATP formation. In uncoupled mitochondria, there will be no change in the amount of oxygen reduced during lactate oxidation, and the GTP formed by the substrate level phosphorylation will be the only high-energy phosphate compound generated. The P/O ratio is 0.2.

   c. The complete oxidation of dihydroxyacetone phosphate (DHAP) is described in the following sequence.

- Triose phosphate isomerase: DHAP → glyceraldehyde-3-phosphate
- Ga3P dehydrogenase: Ga3P + NAD$^+$ + P$_i$ → 1,3-bisphosphoglycerate + NADH
  (NADH, extramitochondrial, P/O = 1.5/1)
- Phosphoglycerokinase: 1,3-bisphosphoglycerate + ADP → 3-phosphoglycerate + ATP
  (P/O = 1/0)
- Phosphoglyceromutase: 3-phosphoglycerate → 2-phosphogylcerate
- Enolase: 2-phosphoglycerate → PEP
- Pyruvate kinase + ADP → pyruvate + ATP (P/O = 1/0)

Oxidation of pyruvate by PDH and TCA cycle: (12.5 ATP/5 oxygen)
Sum is 16 moles of ATP per 6 gram-atoms oxygen reduced. The P/O = 2.7.

   d. In the presence of 2,4-dinitrophenol, there will be 3 moles of ATP formed by substrate-level phosphorylation catalyzed by phosphoglycerokinase, pyruvate kinase, and succinate thiokinase. No ATP is formed by oxidative phosphorylation, but 6-gram-atoms of oxygen are reduced. The P/O ration will be 0.5.

17   Photosynthesis and Other Processes Involving Light

Chapter 17 Answers

2.   Blue light: $\dfrac{64\ \text{Kcal}}{\text{mol photons}}\ \left|\ \dfrac{\text{mol ATP}}{7.5\ \text{Kcal}}\right.$ = 8.5 moles ATP/mol of photons

Red light: $\dfrac{41\ \text{Kcal}}{\text{mol photons}}\ \left|\ \dfrac{\text{mol ATP}}{7.5\ \text{Kcal}}\right.$ = 5.5 moles ATP/mol of photons

4.   a. The relationship between the wavelength and the energy of light is

$$E = \frac{hc}{\lambda}$$

where

E is energy in electron volts

h is Plancks constant: 4.2×10^{-15} eV · s

c is speed of light: 3×10^{10} cm^{-1} s

λ is wavelength: 8.7×10^{-5} cm (870 nm)

$$E = \frac{(4.2 \times 10^{-15}\ \text{eV} \cdot \text{s})(3 \times 10^{10}\ \text{cm} \cdot \text{s}^{-1})}{(8.7 \times 10^{-5}\ \text{cm})}$$

$$E = 1.4\ \text{eV}$$

and

E(kcal/einstein) = nF [E(electron volts)]

where

n is 1 electron-equivalent per einstein and F(Faraday's Constant) is 23.06 kcal V^{-1} eq^{-1}.

E(kcal/einstein) = (1 electron-equivalent einstein)(23.06 kcal V^{-1} eq^{-1})(1.45 eV)

$$E = 33\ \text{kcal/einstein}$$

4.   **(continued)**

    b. The total potential span between the first excited singlet state and the ground state is −1.45 electronvolts as calculated in part (a). Thus absorption of a photon will cause a 1.45 volt separation of an electron for PS_{870} to a more negative (more strongly reducing) potential. Thus:

$$E^{o\prime}_{final} - E^{o\prime}_{initial} \;\; = \;\; -1.45 \text{ V}$$

$$E^{o\prime}_{final} - 0.45 \text{ V} \;\; = \;\; -1.45 \text{ V}$$

$$E^{o\prime}_{final} \;\; = \;\; -1.0 \text{ V}$$

6.   Oxidation of compound A by photosystem II (PS II) in chloroplasts may require the concerted operation of both photosystem I (PS I) and PS II. Inhibition of electron flow between the photosystems would effectively inhibit oxidation of compound A. DCMU is a herbicide that inhibits electron transfer between PS II and PS I. Therefore, if oxidation of compound A requires operation of both photosystem I and photosystem II, oxidation of compound A will be inhibited by DCMU. Moreover, if compound A is oxidized by PS II, the action spectrum for oxidation of the compound will parallel the absorbance spectrum PS II. Oxidation will diminish if chloroplasts are illuminated with red light (>680 nm), a phenomenon sometimes called red drop. However, if compound A is oxidized by photosystem I, DCMU will not inhibit the oxidation nor will the action spectrum of compound A oxidation exhibit red drop.

8.   Oxidation of H_2O to O_2 by plant chloroplasts require the operation of both photosystem I and II, whose reaction center pigments absorb maximally at 700 and 680 nm, respectively. At wavelengths greater than about 700 nm the oxygenic photosystems (PS II) ceases to absorb light and fails to be activated, whereas the PS I absorbs light and is still activated at this wavelength. The quantum yield of O_2 evolution from plants as a function of light absorbed parallels the absorption spectrum until the wavelength of impinging light is greater than approximately 700 nm. Although the chloroplasts absorb wavelengths greater than 700 nm, O_2 evolution declines because the oxygenic photosystem II is no longer activated.

10.   Ribulose bisphosphate carboxylase/oxygenase (RuBisCO) catalyzes the addition of CO_2 to ribulose-1,5-bisphosphate, yielding two moles of 3-phosphoglycerate per mole of CO_2 added. Alternatively, RuBisCO catalyzes the addition of O_2, to ribulose-1,5-bisphosphate, yielding one mole each of 3-phosphoglycerate and phosphoglycolate. Either CO_2 or O_2 binds to the catalytic site of the ribulose bisphosphate carboxylase/oxygenase (RuBisCO) and reacts with the ribulose-1,5,-bisphosphate that is bound to the enzyme. Molecules that bind to the same form of the enzyme (active site) are competitive inhibitors. Thus O_2 competitively inhibits carboxylase activity and CO_2 competitively inhibits oxygenase activity.

12. The mechanism requires an approach of a local pair from the pyruvate keto group onto the alpha phosphorus of the ATP. A hydrogen ion is lost from the pyruvate CH_3 group and PP_i is eliminated from the ATP. The resulting enol pyruvate adenylate is approached by a phosphate, eliminating an AMP and producing phosphoenolpyruvate.

pyruvate   ATP

phosphoenolpyruvate

14. Both electron transport and photosynthesis have cytochromes, quinones, and iron sulfur proteins as components.

16. a. $NADP^+$ is reduced by the $NADP^+$-ferredoxin oxidoreductase whose reductant source is the noncyclic electron flow in the chloroplast. Electrons are transferred to PS II from the oxidation of water via the chloroplast electron-transfer system to PS I. Illumination of PS I generates the strong reductant required to reduce $NADP^+$ to NADPH.

16. **(continued)**

   b. Addition of 3 moles of CO_2 to 3 moles of ribulose-1,5-bisphosphate yields 6 moles of 3-phosphoglycerate. Six moles of ATP are used by phosphoglycerokinase to form 6 moles 1,3-bisphosphoglycerate which are subsequently reduced to glyceraldehyde-3-phosphate (requiring 6 moles NADPH). The 6 moles of glyceraldehyde-3-phosphate are used in the Calvin cycle to regenerate 3 moles of ribulose-5-phosphate with one mole of triose phosphate remaining. Phosphoribulokinase uses 3 moles of ATP to phosphorylate the 3 moles of ribulose-5-phosphate in order to sustain the cycle (in addition to the 9 moles of ATP required, 6 moles of NADPH are used in the formation of 1 moles of triose phosphate).

   c. The ATP consumption by the Calvin cycle in the bundle sheath cells of the C_4 plant does not differ from that of the C_3 plant (9 moles of ATP per mole of triose phosphate). However, the fixation and transfer of CO_2 consumes PEP in the mesophyll cells in the formation of oxaloacetate. Pyruvate is formed by the oxidative decarboxylation of malate in the bundle sheath cells. Resynthesis of PEP from pyruvate is catalyzed by the pyruvate, phosphate dikinase. The products of the reaction are PEP, AMP, and pyrophosphate. The pyrophosphate of AMP to ATP requires two high-energy phosphate equivalents. Thus, two addition ATPs are required in the C_4 plant compared to the C_3 plant.

   d. There is no additional consumption of NADPH in the C_4 pathway of CO_2 fixation. NADPH is used in the mesophyll cell to reduce oxaloacetate to malate, but the NADPH is regenerated upon oxidative decarboxylation of malate to from CO_2 and pyruvate as catalyzed by the $NADP^+$-malic enzyme.

18. The designation C_3 indicates plants whose product of CO_2 fixation is initially 3-phosphoglycerate. C_4 designates plants whose initial product of CO_2 fixation is a 4-carbon product (oxaloacetate). In C_4 plants, the CO_2 is released from the 4-carbon product at a concentration greater than atmospheric levels. Carboxylase activity of RuBisCO in C_4 plants is enhanced relative to carboxylase activity in C_3 plants because of the increased CO_2 concentration and consequent inhibition of the oxygenase activity.

18 Structures and Metabolism of Polysaccharides and Glycoproteins

Chapter 18 Answers

2.

4. Some bacteria synthesize the glucose polymer dextran without the use of nucleotide sugars such as UDP-glucose or ADP-glucose. The (1,6) linkage is catalyzed by dextran sucrase using the disaccharide sucrose as the substrate [n sucrose → dextran + n fructose]. The energy for this reaction comes from the glycosidic bond between glucose and fructose, which is broken to form the linkages between the glucose monomers (this could be called a glycosidic exchange reaction).

6. The large variety of different oligosaccharides results from a limited number of sugars. Sugars have a large number of hydroxyl groups that form glycosidic bonds with the anomeric carbons of other sugars; these anomeica hydroxyl groups can be either α or β conformation. Over eighty different glycosidic linkages are known. With different sugars, number of residues, branching, and other combinations, the number of different oligosaccharides possible is quite large.

8. Many proteins are glycosylated and require that proper oligosaccharides be added for proper folding and translocation. If one of these glycoproteins was expressed in *E. coli*, it would be improperly modified, thus generating a nonfunctional protein. There is much interest in expressing these human glycoproteins in mammalian systems, such as in the milk of transgenic cows.

10. The sequence, bond geometry, and linkage type of monosaccharides in an oligosaccharide or polysaccharide is determined by the specificity of the glycosyltransferases involved (see e.g., fig. 18.22, 18.25, and 18.29).

12. In general amino groups are added to biological molecules in locations once occupied by keto-groups. Because ketoses have carbon-2 as the carbonyl group the majority of the amino sugars in nature are 2-aminosugars. Also, by inspection of figure 18.2 it can be seen that fructose-6-phosphate is the precursor of the amino sugars included in the figure.

14. If carbon one of glucose is oxidized to a carboxylic acid, gluconic acid results. Gluconic acid cannot form a cyclic hemiacetal and cannot form glucosidic bonds which are required for participation in oligo- or polysaccharides.

16. Figure 18.29 is a series of reactions that can be written in a cyclic format with the undecaprenol phosphate (P-Lipid) functioning as a carrier which is acted upon by the series of reactions. The P-Lipid, therefore, does "go" around the cycle.

18. The major problem in the synthesis of complex carbohydrates outside the cell is the lack of an external energy source such as ATP outside the cell. The biosynthesis of a bacterial cell wall such as the peptidoglycan occurs mainly inside the cell, the final step being the cross-linking of the peptidoglycan strands outside the cell. This reaction is a transpeptidation, which does not require any energy source. A peptide bond is broken and another one is formed with the release of alanine. The final step of the synthesis of O-antigens in *Salmonella typhimurium* also occurs outside the cell. In this case the activated forms of the precursors are synthesized inside the cell on a lipid carrier (the same one used in peptidoglycan) and transferred to the outside surface of the cell (held in place by the lipid carrier). The activated galactose of a lipid-linked polymer attacks the mannose of the tetrasaccharide, which is also attached to undecaprenol phosphate (the lipid carrier). In both of these complex reactions to produce extracellular complex carbohydrates a lipid carrier is used to keep intermediates on the surface of the cell. The polymerization is either a transpeptidation (equal bonds broken and formed) or an activated sugar using the lipid carrier.

19   Lipids and Membranes

Chapter 19 Answers

2. Phosphatidylethanolamine would be expected to have the faster rate of flip-flop motion because it is a zwitterion. Phosphatidylserine requires a separate cation (counterion) that would have to diffuse through the membrane at the same rate as the phosphatidylserine.

4. Unless covalent bonds are broken when something penetrates a membrane nothing is "broken." A major factor in membrane formation is the hydrophobic association of the fatty acyl portions of the phospholipids that comprise the membrane. Once the object that penetrates a membrane is removed, unless there has been translocation of the components, the membrane can reform by re-establishment of the hydrophobic interactions and "fusing" the phospholipids.

6. Triacylglycerides are neutral lipids, readily soluble in organic but not aqueous media, and are not amphipathic. Triacylglycerides are stored in adipocytes in a hydrophobic environment, providing a ready source of metabolic energy. Phospholipids are amphipathic because the polar phosphoglyceryl portion of the molecule is soluble in aqueous media, whereas the fatty acids esterified to the glyceryl portion are hydrophobic. Moreover, the phosphate group may be esterified with choline, ethanolamine, or serine, adding to the polar nature of that portion of the molecule. Phospholipids in biological membranes associate in bilayers in which the hydrophobic fatty acyl "tails" form the interior of the bilayer while the polar phosphoryl groups interact with the aqueous medium on either side of the bilayer.

8. Globular proteins found in the cytosol have tertiary structures that allow the hydrophobic amino acid side chains to assume a more thermodynamically favorable association in the interior of the protein because water is excluded. The polar side chains are most frequently found on the surface of the protein in contact with the aqueous environment. Hydration of the polar groups and the diminished organization of water upon removal of the nonpolar side chains from aqueous contact stabilize the water-soluble protein.

    Integral membrane proteins have a large portion of the surface "dissolved" in the hydrophobic lipid bilayer of the membrane. In this case, exterior polar groups would be less stable then exterior hydrophobic groups.

    The hydrophobic amino acid side chains on the exterior of the integral membrane protein are stable in the water-free environment of the hydrophobic membrane interior. The polar side chains must be inside the protein, associated with water or in ionic bonds with other polar side chains. The portion of the membrane protein protruding from the membrane and n contact with the aqueous environment is composed primarily of polar amino acids, similar to soluble proteins.

10. Peripheral proteins are bound to the inner or outer aspects of the membrane through weak ionic interactions that include association with phospholipid head groups, by electrostatic or ionic interaction with a hydrophilic region of an integral membrane protein or through divalent metal ion bridging to the membrane surface. Integral proteins are dissolved into the lipid bilayer of the membrane through interactions of the hydrophobic amino acid side chains and fatty acyl groups of phospholipids. These interactions exclude the hydrophobic residues from aqueous contact.

    Peripherally bound proteins may be released without disrupting the membrane. Thus, increased salt concentration shields ionic charges and weakens the charge—charge interactions between the peripheral protein and the membrane components. Although increased ionic strength may weaken the salt bridges contributed by divalent metal ions, application of a chelator to sequester the divalent ions may promote release of the peripheral protein. Changing the relative proportion of protonated/deprotonated groups by adjusting pH would, in principle, affect binding of peripheral proteins to the membrane.

    None of the conditions described would be expected to release integral membrane proteins. For example, high ionic strength fosters, rather than weakens, hydrophobic interactions that bind integral proteins to the membrane. In order to remove integral membrane proteins, the membrane must be disrupted by addition of detergents or other chaotropic reagents to solubilize the protein and to prevent aggregation and precipitation of the hydrophobic proteins upon their removal from the membrane. For example, succinate dehydrogenase, a membrane-bound primary dehydrogenase, is an integral protein in the inner mitochondrial membrane and is removed only upon dissolution of the membrane with caotropic reagents.

12. The amino acid side chains in an α-helix protrude from the axis of the helix and interact either with solvent or with other amino acid side chains (in a folded protein). In the α-helix of single-span integral protein, each amino acid side chain would interact with the hydrophobic interior of the membrane. In principle, it is energetically unfavorable to place a hydrophilic residue in a nonaqueous environment. In integral proteins with multiple α helices that span the membrane, hydrophilic side chains from different helical segments may interact and in some cases form a channel through which ions may diffuse. Portions of the helical segments exposed to the lipid will contain primarily hydrophobic amino acid residues.

14. Phosphatidylcholine describes a class of molecules consisting of *sn*-glycerol-3-phosphate linked through a phosphodiester bond to the —OH group of choline (see the structure shown here). This portion of the molecule is common to all the phosphatidylcholines. The fatty acids esterfied to the C-1 and C-2 OH groups of the glyceryl portion of the molecule (R_1 and R_2 in the structure) provide the diversity of phosphatidylcholine structures.

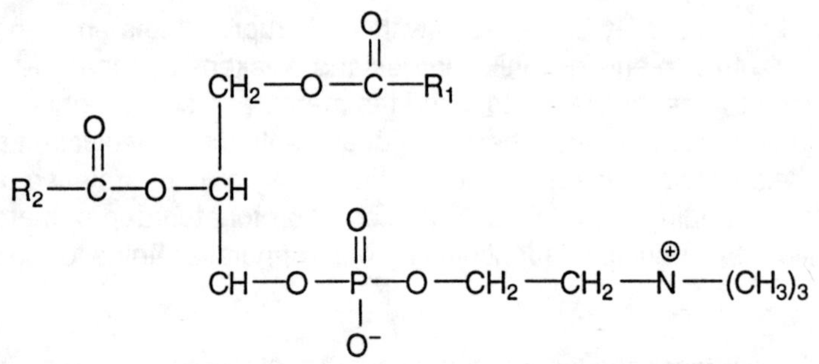

6. Which of the following statements is true about the enzyme phosphorylase?
    a. Phosphorylase is phosphorylated on a threonine residue.
    b. Phosphorylase is phosphorylated on a serine residue.
    c. Phosphorylase is phosphorylated on a tyrosine residue.
    d. Phosphorylase is not phosphorylated.
    e. Phosphorylase is phosphorylated on a tryptophan residue.
    answer: b

7. Which of the following statements is true about the enzyme phosphorylase b?
    a. Phosphorylase b is phosphorylated on a threonine residue.
    b. Phosphorylase b is phosphorylated on a serine residue.
    c. Phosphorylase b is phosphorylated on a tyrosine residue.
    d. Phosphorylase b is not phosphorylated .
    e. Phosphorylase b is phosphorylated on a tryptophan residue.
    answer: d

8. Which of the following statements is true for allosteric enzymes?
    a. The substrate binds more tightly to the T form.
    b. The substrate does not bind to the T form.
    c. The binding of the substrate favors the transition of the T form to the R form.
    d. The binding of the substrate favors the transition of the R form to the T form.
    e. None of the above.
    answer: c

9. Which of the configurations for a tetrameric enzyme does the symmetry model allow?
    a. T T T T
    b. T R R T
    c. R T T R
    d. all of the above
    e. none of the above
    answer: a

10. Which of the following statements is true about allosteric enzymes?
    a. Allosteric inhibitors bind preferentially to the R form.
    b. Allosteric inhibitors bind preferentially to the T form.
    c. Allosteric inhibitors bind equally well to T and R forms.
    d. Allosteric inhibitors do not bind to the T form.
    e. None of the above.
    answer: b

4. c. (continued)

Solving for X gives 14.3 nM. Therefore, at equilibrium,

Side 1: 64.3 mM K^+, 50 mM Na^+, 144.3 mM Cl^-

Side 2: 85.7 nM K^+, 85.7 mM Cl^-

To calculate $\Delta\psi$, one can use either the K^+ or Cl^- gradient:

$$\Delta\Psi = -60 \log \frac{64.3}{85.7} = +60 \log \frac{114}{85.7}$$

$$= 60(0.125) = 7.5 \text{ mV}$$

6. (This potential is such that side 1 is positive and side 2 is negative.)

This problem could be approached in a number of ways, but the simplest of these will be described for illustrative purposes.

a. If sucrose transport were energized via H^+ symport, membrane vesicles should transport sucrose in the presence of an electron donor, such as *D*-lactate, which would set up a ΔP. Proton ionophores should inhibit sucrose uptake under these conditions.

b. If Na^+ symport were involved, active uptake should be absolutely dependent on extravesicular (or extracellular) Na^+, simulated by a $\Delta\Psi$ (interior negative) and insensitive to ΔpH (at constant $\Delta\Psi$).

c. If there were a sucrose binding protein necessary for transport, cells subjected to cold osmotic shock, spheroplasts, and plasma membrane vesicles should be all defective in sucrose transport regardless of the energy source tested.

d. If a PTS for sucrose were present, crude extracts of cells should phosphorylate sucrose by a reaction dependent on PEP, but not on ATP.

8. a. Rearranging equation (5) from the text, we have

$$D = \frac{V \cdot 1}{[So] - [So']}$$

V for diffusion across a membrane has the dimensions of number of molecules per unit area per unit time [e.g., mole/(cm^2 · s)]. Therefore,

$$\text{Dimensions of } D = \frac{\left(\dfrac{\text{mole}}{\text{cm}^2 \cdot \text{s}}\right) \cdot (\text{cm})}{\left(\dfrac{\text{mole}}{\text{cm}^3}\right)} = \frac{\text{cm}^4}{\text{cm}^2 \cdot \text{s}} = \frac{\text{cm}^2}{\text{s}}$$

8. **(continued)**

   b. From the data given, the area of the cross section of a porin pore is 0.78×10^{-14} cm². Fifty molecules of glucose per second equals 0.83×10^{-22} moles per second. Therefore,

   $$D = \left(-\frac{0.83 \times 10^{-22} \text{ mole}}{0.78 \times 10^{-14} \text{ cm}^2} \right) \cdot \left(\frac{4 \times 10^{-7} \text{ cm}}{3 \times 10^{-9} \text{ mole/cm}^3} \right)$$

   $$= 1.4 \times 10^{-6} \frac{\text{cm}^2}{\text{s}}$$

   To be more precise, the diameter of a glucose molecule is about half the diameter of a porin pore (0.5×10^{-9} m). Thus the effective pore diameter is only about 0.5×10^{-9} m if we subtract the diameter of glucose. The effective cross-sectional area for diffusion is actually 0.19×10^{-14} cm². Taking this into account,

   $$D = 5.7 \times 10^{-6} \frac{\text{cm}^2}{\text{s}}$$

   This value is close to the diffusion coefficient for glucose is dilute aqueous solution, which is about 7×10^{-6} cm²/s at room temperature and suggests that glucose is more or less freely diffusible in the pore channel.

10. Consider a typical "carrier" that has its single substrate binding site exposed to either the inside or the outside of the cell but never to both simultaneously. Efflux of substrate from inside to outside would involve a conformational change in the protein, switching the binding site accordingly. Initially, in the absence of added substrate to the outside, the concentration of substrate outside the cell or vesicle would be negligible. Since the return of the carrier to the "inside" conformation would be necessary for it to participate in another efflux step, addition of nonradioactive substrate to the outside should stimulate efflux by increasing the rate of switching of the reversible carriers back to the "inside" conformation. In other words, the transport-associated conformational change in either direction is accelerated by substrate binding.

12. The effect of Hg^{2+} on the transport of D-glucose is to denature the protein transporter leaving only the passive transport of glucose. This evidence also supports the experiment in problem 9 that there is carrier-mediated transport of glucose in these cells and that the transporter is protein. Since the rate of transport of D-glucose in the mercury-treated cells is the same as L-glucose, this slow rate of transport is not carrier-mediated but due to simple diffusion through the cell membrane.

14. Although there are many seemingly different permeases, they all seem to have a hydrophilic channel (pore) that is gated. The translocation involves a conformational change in the protein that in facilitated diffusion serves to equilibrate the concentrations of the transport substrate on both sides of the membrane. In active transport, the binding affinity of the transport substrate is higher outside than inside the cell. These common features of many permeases is consistent with the evolutionary relatedness of all transport systems found in cells and can allow them to be described by a unified model.

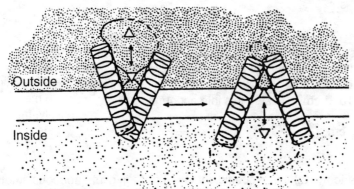

(a) Facilitated diffusion

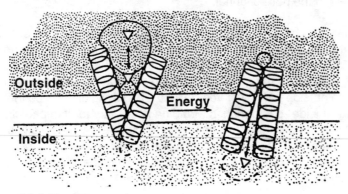

(b) Active transport

21　Metabolism of Fatty Acids

Chapter 21 Answers

2. Using calculations based on figures 21.4 and 21.6 and assuming NADPH = NADH, arachidic produces 134 moles of ATP/moles FA while arachidonic gives 126 moles of ATP/mole of fatty acid. This differences in ATPs is rather minor. It is unlikely, therefore, that any biological difference caused by dietary saturated vs. unsaturated fats is due to their ATP yields.

Arachidic acid		ATP's
beta-oxidation produces 10 AcSCoA	10×10	= 100
9 FADH$_2$ + 9 NADH	9×4	= 36
fatty acid activation		-2
sum		134

Arachidonic acid (see figure 21.6)		
beta-oxidation produces 10 AcSCoA	10×10	= 100
7 FADH$_2$ + 9 NADH − 2 NADPH =	$7 \times 1.5 + 7 \times 2.5$	= 28
fatty acid activation		-2
sum		126

4. a. | Glucose | 2 cytoplasmic NADH | 3 |
    |---|---|---|
    | | 2 cytoplasmic ATP | 2 |
    | | 2 NADH from pyr deH$_2$ase | 5 |
    | | 2 AcSCoA | 20 |
    | | sum | 30 ATP |

    115

   b. 64 moles ATP per mole decanoic acid

   c.
$$\frac{30\ \text{ATP}}{\text{glucose}} \left| \frac{-7.5\ \text{Kcal}}{\text{ATP}} \right| \frac{\text{Glucose}}{-669.3\ \text{Kcal}} \left| \frac{100}{} \right| = 33\%$$

$$\frac{64\ \text{ATP}}{\text{C10 acid}} \left| \frac{-7.5\ \text{Kcal}}{\text{ATP}} \right| \frac{\text{C10 Acid}}{-1452\ \text{Kcal}} \left| \frac{100}{} \right| = 32\%$$

   d. $\dfrac{9\ \text{kcal/g fat}}{4\ \text{kcal/g carbo}} = 2.25 \qquad \dfrac{64\ \text{ATP/fat}}{30\ \text{ATP/glu}} = 2.13$

    The ratios are quite close.

6. The fatty acids are transported across the mitochondrial inner membrane as fatty acyl carnitine esters. The fatty acid is activated for metabolism as the fatty acyl-CoA derivative in the cytoplasm, a process requiring the equivalent of 2 moles ATP. Neither the free fatty acid nor the fatty acyl-CoA can freely traverse the mitochondrial inner membrane. Carnitine acyl transferase associated with the outside of the mitochondrial inner membrane (CAT I) catalyzes transfer of the fatty acyl group to carnitine. The fatty acyl carnitine ester is translocated across the inner membrane in exchange for carnitine. A carnitine acyl transferase on the matrix aspect of the inner membrane (CAT II) catalyzes the transfer of the acyl group to CoASH, reforming the fatty acyl-CoA. The result is translocation of the fatty acid from cytoplasm to mitochondrial matrix with preservation of the active thioester. The acyl-CoA is then available for β-oxidation in the mitochondrial matrix.

8. If the following simplifications are considered (1) the transfer of materials across membranes is ignored and (2) NADH is considered to be equal to NADPH, there will be an energy consumption in the fatty acid to acetyl-CoA to fatty acid sequence.

    For beta oxidation to the acetyl-CoA level and where n = number of carbon atoms in the fatty acid:

    ATP production = $4((n/2) - 1) - 2$

    The number of ATP required for fatty acid biosynthesis:

    ATP = $((n/2) - 1)(2) + ((n/2) - 1)(5)$
    mal-CoA syn   NADPH required
    citrate lyase   for reduction

    Combining beta oxidation and biosynthesis:

    ATP = $4((n/2) - 1) - 2 - [((n/2) - 1)(2) + ((n/2) - 1)(5)] = 3(n/2) - 3$

    This scheme represents a net loss (consumption) of ATP, therefore, there should be a weight loss.

10. Beta ketoacids readily decarboxylate by way of a resonance stabilized carbanion.

12. Carnitine is the cofactor required to transport long-chain fatty acids across the mitochondrial inner membrane to the matrix for β-oxidation. Carnitine deficiency would limit fatty acid transport to the mitochondrial matrix and limit energy conversion (as ATP) from the β-oxidation pathway. Glucose would thus be used as an energy source both in the fed and fasted state because of the limited fatty acyl oxidation.

Gluconeogenesis requires both a carbon source (e.g., lactate, pyruvate, or alanine but not acetyl-CoA) and a supply of ATP. Hepatocytes utilize fatty acid oxidation to supply ATP for gluconeogensis. Without the energy supply, gluconeogenesis fails, and the blood glucose level fails (hypoglycemia). A plausible link between carnitine deficiency and hypoglycemia is suggested based on energy supply.

14. Ketone bodies (β-hydroxybutyrate and acetoacetate) are synthesized by the liver-specific mitochondrial HMG-CoA synthase and HMG-CoA lyase and are transported in the blood to nonhepatic tissue. In nonhepatic tissues, β-hydroxybutyrate is oxidized to acetoacetate and subsequently activated to acetoacetyl-CoA. Acetoacetyl-CoA formation is catalized by β-oxyacid-CoA-transferase, an enzyme that transfers the coenzyme A group from succinyl-CoA to acetoacetate. If the transferase activity were markedly diminished, the rate of ketone body oxidation in nonhepatic tissue would be decreased, but the rate of ketone body formation in the liver likely would not diminish. Thus, there would be an increase in the acetoacetate and β-hydroxybutyrate concentration in the bloodstream. If the transferase were absent, ketone body oxidation in nonhepatic tissues would be abolished. The blood concentration of the ketone bodies would soar, particularly when the glucagon/insulin ration increased, signaling increased lipid oxidation in the liver. Ketonuria and acidemia likely would result. In either case, the nonhepatic tissues could gain little or no metabolic energy from ketone body oxidation and would have greater dependency on glucose as a metabolic energy source. Glucose oxidation rates in those tissues would increase.

16. The fatty acid synthase in mammalian liver is a multifunctional protein and may be divided into three domains: I, substrate entry and condensation; II, reduction and dehydration; and III, hydrolytic release of product. All of the enzymatic activities required for the synthesis of palmitate from acetyl and malonyl groups reside on the multifunctional protein.

    Two sulfhydryl groups, one from cysteine in the β-ketoacyl-ACP synthase and the other from phosphopantetheinyl group of the acyl carrier portion of the synthase, accept the acetyl and malonyl groups, which are condensed and remain bound to the acyl carrier portion of the protein. The ketoacyl group must be transported, in succession, to the keto reductase, hydroxyaceyl dehydratase, and enoyl reductase active sites on the protein, but remain bound to the acyl carrier.

    The structural model proposed for the fatty acid synthase is a cylinder approximately 16 nm long, with domain I at the N terminal, domain II midway between the N and C terminals, and domain III at the C terminal. The acyl group attached to the flexible pantothenate on the acyl carrier apparently cannot physically reach, and thus cannot chemically interact, with domain I from the same multifunctional protein. However, the fatty acyl synthase proteins dimerize so that the N terminal (domain I) of one protein is proximal to the C terminal (domains II and III) of the other protein. Each protein functionally complements the other (see Wakil, S., *Biochem.* 28 4523–4530 (1989)).

18. a. Citrate is formed from the condensation of acetyl-CoA and oxaloacetate in the mitochondrial matrix and is either oxidized by the TCA cycle enzymes or is exported to the cytosol to supply substrate for fatty acid biosynthesis. When metabolite supply to the mitochondria is elevated or in excess, and energy supply is being met, the excess metabolite, as citrate, is transported to the cytosol. The citrate is cleaved to acetyl-CoA and oxaloacetate by the ATP-dependent citrate lyase. Acetyl-CoA is the carbon source for palmitate synthesis.

      Citrate plays a regulatory role in fatty acid synthesis as a potent activator of the acetyl-CoA carboxylase. Upon binding citrate, the liver acetyl-CoA carboxylase polymerizes to an aggregate (approximately 4–8×10^6 M_t) with a concomitant increase in catalytic activity. The rate of carboxylation of acetyl-CoA to form malonyl-CoA increases, as does palmitate synthesis. Thus, when energy and metabolite supply is increased, citrate is a source of substrate and an activator of fatty acid synthesis.

18. (continued)

b.

(A)

$$Citrate_{mitochondrial} \rightarrow citrate_{cytoplasmic}$$

(B)

$$Citrate_{cytoplasmic} + ATP + CoASH \rightarrow acetyl\text{-}CoA + OAA$$

(C)

$$OAA + NADH \rightarrow L\text{-malate} + NAD^+$$

(D)

$$L\text{-malate} + NADP^+ \rightarrow pyruvate + CO_2 + NADPH$$

(E)

$$Pyruvate_{cytoplasmic} \rightarrow pyruvate_{mitochondrial}$$

(A)   Citrate transport system
(B)   ATP-dependent citrate lyase
(C)   Cytoplasmic malate dehydrogenase
(D)   NADP$^+$-malic enzyme
(E)   Pyruvate transporter

Citrate supplies acetyl-CoA and reducing equivalents (NADPH) for fatty acid synthesis.

Alternatively, L-malate may be returned to the mitochondrial matrix via the malate shuttle, bypassing reactions (D) and (E).

20. a. Both the β-oxidation of fatty acids and palmitate synthesis occur in the liver cell. The oxidation and synthesis of fatty acids, although mechanistically different, may be considered metabolically opposing pathways and, if not regulated, could result in depletion of cellular ATP and reductant levels. Were the pathways in the same cellular compartment, the rate-limiting step of each pathway would likely be coordinately regulated. Recall the activation of phosphofructokinase-1 (glycolysis) and inhibition of fructose bisphosphatase-1 (gluconeogenesis) by the fructose-2,6-bisphosphate.

Another strategy used by the cell to regulate metabolism in opposing pathways is separation of the pathways into different cellular compartments. Thus, regulation of metabolite supply may then limit activity of the pathway. Enzymes catalyzing β-oxidation of fatty acids are mitochondrial; those synthesizing palmitate are cytoplasmic.

b. The carnitine acyltransferase associated with the outside of the inner mitochondrial membrane (CAT I) is inhibited by malonyl-CoA with an inhibition constant (K_i) of approximately 1 to 2 micromolar (see McGarry, J.D., and D.W. Foster, *Annu. Rev. Biochem* 49:395–420 (1980)). Inhibition of the carnitine acyltransferase effectively prevents translocation of the long-chain fatty acid from the cytoplasm into the mitochondrial matrix, the site of β-oxidation. An inhibitor of CAT I would be expected to exert the same inhibitory effect as malonyl-CoA on β-oxidation.

22   Biosynthesis of Membrane Lipids

Chapter 22 Answers

2. A possible explanation includes: serine condenses with pyridoxal phosphate, the serine loses a CO_2, a carbanion remains on what was the alpha carbon, the carbanion approaches the carbonyl carbon of palmitoyl-CoA, and CoASH is eliminated. Following the hydrolysis of the imine and release of pyridoxal phosphate, 3-ketosphinganine is formed.

4. If condensation of the serine residue preceded decarboxylation (see problem 22.2) the carbanion (i.e., required for approach on the carbonyl of palmitoyl-CoA) would have to come from the loss of the alpha hydrogen. Initial decarboxylation does not require loss of the alpha hydrogen. In fact, pyridoxal phosphate based decarboxylases are known to retain the alpha hydrogen.

6. Phosphatidylserine decarboxylase is a pyridoxal phosphate enzyme. This is expected because phosphatidylserine has a free amino group and typically amino acid decarboxylyases are pyridoxal phosphate enzymes.

8. For this couple, 25% of their children would have the disease, 50% would be carriers, and 25% would be normal (disease free and not carriers).

10.

phosphatidase

12. Phospholipids are very important to cells because they are the major components of membrane structure. The demand for phospholipids is highest in growing cells, such as those in a developing fetus. Any genetic defects in the biosynthesis of phospholipids would be lethal in the early stages of fetal development.

14. Deacylation-reacylation of phospholipids occurs in various tissues at both the SN-1 and SN-2 positions. This remodeling allows alternation of the phospholipid without having to completely synthesize the entire molecule. In lung tissue the surfactant dipalmitoylphosphatidlycholine is synthesized by removing the fatty acid in the SN-2 position and then acylating with palmitoyl-CoA. Another example of remodeling occurs when the fatty acid at the SN-2 position is replaced with arachidonic acid. The arachidonic acid is stored there until it is needed for eicosanoid biosynthesis.

16. Ether-linked phospholipids are widely found in nature. Ether-linked lipids are the major component in archaebacteria cell membranes. Eukaryotic membranes contain significant amounts of ether-linked glycerolphospholipids. For example, 50% of all phospholipids in heart tissue are ether-linked in the SN-1 position. Plasmalogens are the major form of ether-linked phospholipids found in the heart and are characterized by a vinyl ether linkage. Alkenyl-ether-containing phospholipids protect cells against singlet oxygen, which can kill cells. Some ether-linked phospholipids are bioactive, such as 1-alkyl-2-acetylglycerolphosphocholine (platelet-activating factor), which reduces blood pressure and causes blood platelets to aggregate.

18. Low doses of aspirin appear to help prevent heart attacks and strokes in people over 40 years old. TXA2 is made in blood platelets and causes platelet aggregation; it is a vasoconstrictor. If the cyclooxygenase is inactivated by aspirin, no new enzyme can be made cause these cells have no nucleus. New platelets would have to be made to regain the synthesis of TXA2. This would take days. Vascular endothelial cells make PGI2, a vasodilator that inhibits platelet aggregation. These cells in the arterial walls have a nucleus and can make more cyclooxygenase in a few hours. If high dose aspirin is taken, the synthesis of PGI_2 does not recover rapidly. Thus, PGI_2 helps prevent heart attacks and strokes. A small amount of aspirin is certainly better in this case.

23   Metabolism of Cholesterol

Chapter 23 Answers

2. Of cholesterol's 27 carbons, 12 originate as the carboxyl (carbon numbers 2, 4, 6, 8, 10, 11, 12, 14, 16, 20, 23, 25) and 15 as the methyl carbon of acetate (carbon numbers 1, 3, 5, 7, 9, 13, 15, 17, 18, 19, 21, 22, 24, 26, 27). See Figure 19.6 for carbon numbering schemes.

4. Typically the fatty acyl group on carbon 2 of phosphatidyl choline is unsaturated. Since this fatty acyl residue is transferred, the fatty acyl group on cholesterol ester is expected to be unsaturated.

6. No. Some examples of other types of intermediates involved in amide bond formation include: a) mixed anhydrides, in peptidoglycans (chapter 18), glutamine (chapter 24) and citrulline (chapter 25) synthesis, b) esters, amino acyl-tRNAs in peptidoglycan synthesis (chapter 18) and protein synthesis (chapter 34).

8. Because of the greater energy content of the thioester bond (vs. an oxygen ester) it is anticipated that the equilibrium of the reaction promoted by CoA: cholesterol acyltransferase would be more in favor of the cholesterol ester.

10. The low plasma cholesterol level exhibited by the defective rats may have any of a number of possible explanations. The low activity of HMG-CoA reductase (a microsomal enzyme) in microsomal fractions prepared from the livers of these rates indicates that a defect in this enzyme is responsible. HMG-CoA reductase catalyzes the rate-limiting step in cholesterol biosynthesis, and a defect in the activity of this enzyme would have sever consequences for the plasma cholesterol level.

    The low activity of HMG-CoA reductase observed for the unusual rats might result from a defect in either the amount or activity of this enzyme. The restoration of activity observed upon addition of normal rat liver cytosolic fractions to the inactive microsomal fractions prepared from the defective rats indicates that the deficiency derives from inactivation of an otherwise normal enzyme. The activity HMG-CoA reductase is known to be subject to regulation by a phosphorylation-dephosphorylation system as shown. The aberrant inactivation of reductase and its reactivation by a normal cytosolic fraction may be explained by a deficiency in this system in either reductase kinase or reductase phosphatase activities (both cytosolic), preventing dephosphorylation and reactivation

    of inactive HMG-CoA reductase- .

10. (continued)

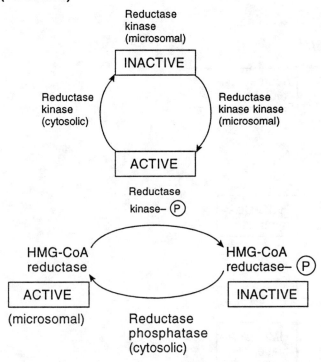

Further support for this as a possible explanation of the low activity of reductase in the unusual rats might be obtained by determining the effect on activity of inactive microsomal fractions, of addition of pure phosphorylase a-phosphatase, the enzyme in rat liver believed to function in these dephosphorylation reactions.

12. Essentially any attempt at an explanation requires evolution concepts. The increasing involvement of membranes, organelles, and organs during evolution combined with the evolutionary development of new metabolic sequences from previously developed enzymatic mechanisms has given us what appears to be a diverse collection of chemical components and processes. Notice within this chapter, for example, what diverse roles nature has found for the steroid ring system.

14. Individuals in an acute diabetic condition general accumulate a considerable quantity of the ketone bodies acetoacetate and β-hydroxybutyrate in plasma (ketonemia) owing to overproduction of acetyl-CoA as a product of fatty acid oxidation in the liver. The synthesis of these ketone bodies, which in the liver occurs by the pathway shown, requires the activity of four principle enzymes: acetoacetyl-CoA thiolase, HMG-CoA synthase, HMG-CoA lyase, and D-β-hydroxybutyrate dehydrogenase.

14. (continued)

The observation that an individual in diabetic shock shows no sign of ketone bodies in the plasma would indicate that one or more of the activities necessary for the production of acetoacetate (acetoacetyl-CoA thiolase, HMG-CoA synthase, and HMG-CoA lyase) must be defective in the individual.

14. **(continued)**
    Cholesterol biosynthesis is known to require the activities of two enzymes common to the pathway for ketone body synthesis. These are acetoacetyl-CoA thiolase and HMG synthase. On the face of it, therefore, it would seem that if cholesterol biosynthesis in the individual were normal, a defect in the activities of either of these two enzymes would be precluded. However, two facts must be borne in mind when considering these observations:

    (1) Cholesterol biosynthesis is a cytosolic process, whereas the synthesis of ketone bodies is confined to the mitochondrial matrix.

    (2) The enzyme activities common to the two pathways are consigned to different isozymes.

    Because of these qualifications, it is not possible to infer from the ability of the individual to synthesize cholesterol normally that the two enzymes activities shared by these pathways are not the site of the defect in ketone body synthesis.

16. The enzyme lecithin: cholesterol acyltransferase (LACT) plays an extremely important role in the transport of cholesterol in plasma. It performs this role in association with one of the longer-lived lipoprotein carriers, HDL. In normal individuals, dietary cholesterol transferred from chylomicrons, and waste cholesterol released into plasma from extrahepatic cells, absorb to HDL and are converted by the action of LACT to cholesterol esters. These esters are then shuttled to other lipoprotein carriers, principally VLDLs and LDLs, destined for conversions to LDLs and subsequent uptake by extrahepatic tissues. Much of the cholesterol ester in human LDLs is cholesterol that has been recycled from tissues in this way.

    In LCAT deficiency, the flow of cholesterol from dietary and waste endogenous sources to extrahepatic tissues is severely impaired. HDL particles, which leave the liver and intestine essentially devoid of cholesterol esters, remain so throughout their lifetime. Other lipoprotein carriers, notable VLDLs and LDLs, suffer a similarly reduced content of cholesterol esters, while plasma exhibits an elevated level of free cholesterol.

24 Amino Acid Biosynthesis and Nitrogen Fixation in Plants and Microorganisms

Chapter 24 Answers

2. The products of asparagine synthetase (Figure 24.15) include AMP and PPi, which indicate an adenylate intermediate. When the nitrogen of the glutamine amide displaces the AMP of the adenylate intermediate the AMP carries with it one of the two original aspartate γ carboxylate oxygens. The second γ carboxylate oxygen becomes the asparagine amide oxygen. Assuming no isotope effect AMP and Asn would each receive one half of the original labeled oxygen.

4. The removal of the phosphate from O-phosphoserine is by a hydrolysis (hydrolase) mechanism, while the removal of the phosphate from O-phosphohomoserine is by an elimination reaction in which a double bond is formed (lyase).

6. The biosynthetic pathways for the majority of the amino acids involve the formation of the corresponding α-ketoacid. Frequently, as the last step in the pathway, a transamination (involving glutamate) produces the amino acid.

8. Glutamate synthetase (figure 24.3) converts a molecule of α-ketoglutarate and glutamine into two molecules of glutamate. The two molecules of glutamate are then turned into two glutamines. One of the glutamines feeds back into the glutamate synthetase step.

10. At some point in the reaction mechanism there is an imine (N = C) intermediate. To produce the amine the imine must be reduced and NADPH is the reducing agent.

12. 2-Aminobutanoate would be expected to be produced from the transamination of α-ketobutanoate which in turn could be derived from the deamination of threonine which is a member of the aspartate family of amino acids.

2-aminobutanoate

14. There are several situations in which an amino acid could be considered to be in two families, e.g., isoleucine, tryptophan, and lysine. Isoleucine is in the pyruvate and aspartate families, tryptophan is in the aromatic and serine pathway, lysine is in the pyruvate and aspartate families while fungal lysine is in the glutamate family.

16. Carbons 1 and 2 of ribose are incorporated into tryptophan (figure 24.26).

18. Glutamine synthase is allosterically inhibited upon binding each of several metabolites to specific sites on the enzyme. The regulation requires that the structure of an inhibitory metabolite bind to a specific portion of the protein in a process not unlike that of substrate binding to the active site of an enzyme. Molecules whose structures are similar may bind to the same regulatory site, but molecules whose structures are dissimilar likely require separate binding sites. Thus, the structures of the diverse regulatory metabolites require different binding sites on the glutamine synthase and are required to explain cumulative inhibition.

20. Tryptophan synthase catalyzes the cleavage of the substrate indolylglycerol phosphate, yielding the products glyceraldehyde-3-phosphate and indole. Whereas glyceraldehyde-3-phosphate is released, indole is retained at the active site and replaces the hydroxyl group of the substrate serine used to form tryptophan. Indole is both formed and used at the synthase active site but is not released and may be considered an enzyme-bound intermediate in synthesis.

22. Mutants in which a specific enzyme is missing, synthesized in low quantity, or synthesized with altered kinetic properties, in principle should accumulate the substrate of the defective enzyme. If the mutant cells were supplied with the end product of the biosynthetic pathway, the committed step to the pathway likely would be inhibited, or synthesis of enzymes of the pathway would be repressed. No accumulation of substrate for the altered enzyme would be observed. Upon consumption of the end product of the pathway, feedback inhibition is alleviated and synthesis of enzymes required in the biosynthetic pathway is derepressed. The activity of each of the enzymes in the pathway is needed to supply the required end product. Were the activity of an enzyme missing or severely curtailed, we would anticipate accumulation of metabolite (or some derivative of the metabolite) on the substrate side of the altered enzyme in the metabolic pathway.

25 Amino Acid Metabolism in Vertebrates

Chapter 25 Answers

2. The average number of enzymes involved in the biosynthetic pathways to the essential amino acids is substantially greater than in the nonessential amino acids. If the diet of our ancestors, throughout evolution, consistently contained an adequate supply of amino acids (i.e., no deleterious effects of an amino acid becoming essential) then the probability of an amino acid becoming essential would relate to the probability of a mutation(s) occurring in a gene(s) encoding that pathway. The greater the number of genes involved in the pathway (assuming genes of equal length) the greater that the amino acid produced by that pathway would become essential.

4. Most transaminases use alpha-ketoglutarate as a substrate and the transamination is intermolecular. The anticipated intermediate would be gama-keto-glutarate-semialdehyde.

6. Start with the ATP promoted phosphorylation of the enol tautamer of 5-oxoproline, allow the imine to be hydrolized to produce glutamyl-gama-phosphate, followed by hydrolysis of the mixed anhydride. With the expenditure of energy (ATP hydrolysis) the cell allows the concentration of 5-oxoproline to be decreased to minimal amounts, which in turn "pulls" the previous reaction.

8. The *N*-acetyl groups in the *de novo* pathway prevent the spontaneous cyclization of glutamate-δ-semialdehyde. Acetylation provides a means of differentiating, for control purposes, similar chemical species involved in the *de novo* arginine pathway from those involved in the salvage pathway (figure 25.2) and proline biosynthesis (figure 24.9). Also, the intermediates are distinct for control purposes.

10. One of the nitrogenous enters the urea cycle via carbonyl phosphate which comes from ammonia (figure 25.7). The second urea nitrogen enters the urea cycle as part of aspartic acid which in turn can come from transamination of oxaloacetate. Glutamate is the source of the amino group in the transamination. Glutamate in turn can arise from α-glutarate by transamination or a reductive amination (glutamate dehydrogenase) involving ammonia.

12. Because of the importance of the urea cycle the capacity to convert into arginine is obvious. Complete loss of the ability to produce ornithine, (a catalyst or carrier in the urea cycle) would limit the organism's control over production of its nitrogen waste product.

14. Based on figure 25.11, Thr, Ala, Ser, Gly, Cys, Asn, Asp, Gln, Glu, His, Arg, Pro, Val, and Met are glucogenic; Lys, Trp, and Leu are ketogenic; and Phe, Tyr, and ILe are both ketogenic and glucogenic.

16. A homozygotic defect in a gene for a protein in glycolysis, the citric acid cycle, or electron transport would probably lead to the death of the cell(s) soon after fertilization because of their critical role in ATP production.

18. Pyridoxal phosphate forms a Schiff base (imine) with the glycine. A carbon bound hydrogen is labile and the resulting carbanion stabilized by resonance back into the pyridoxal phosphate. The carbonanion approaches the carbonyl carbon of the succinyl-CoA. Following the elimination of the CoASH the intermediate shown in figure 25.13 is formed. The intermediate then loses a CO_2 forming a carbanion that is resonance stabilized back into the pyridoxal phosphate. Protonation of the latter intermediate produces δ-aminolevulinate.

18. (continued)

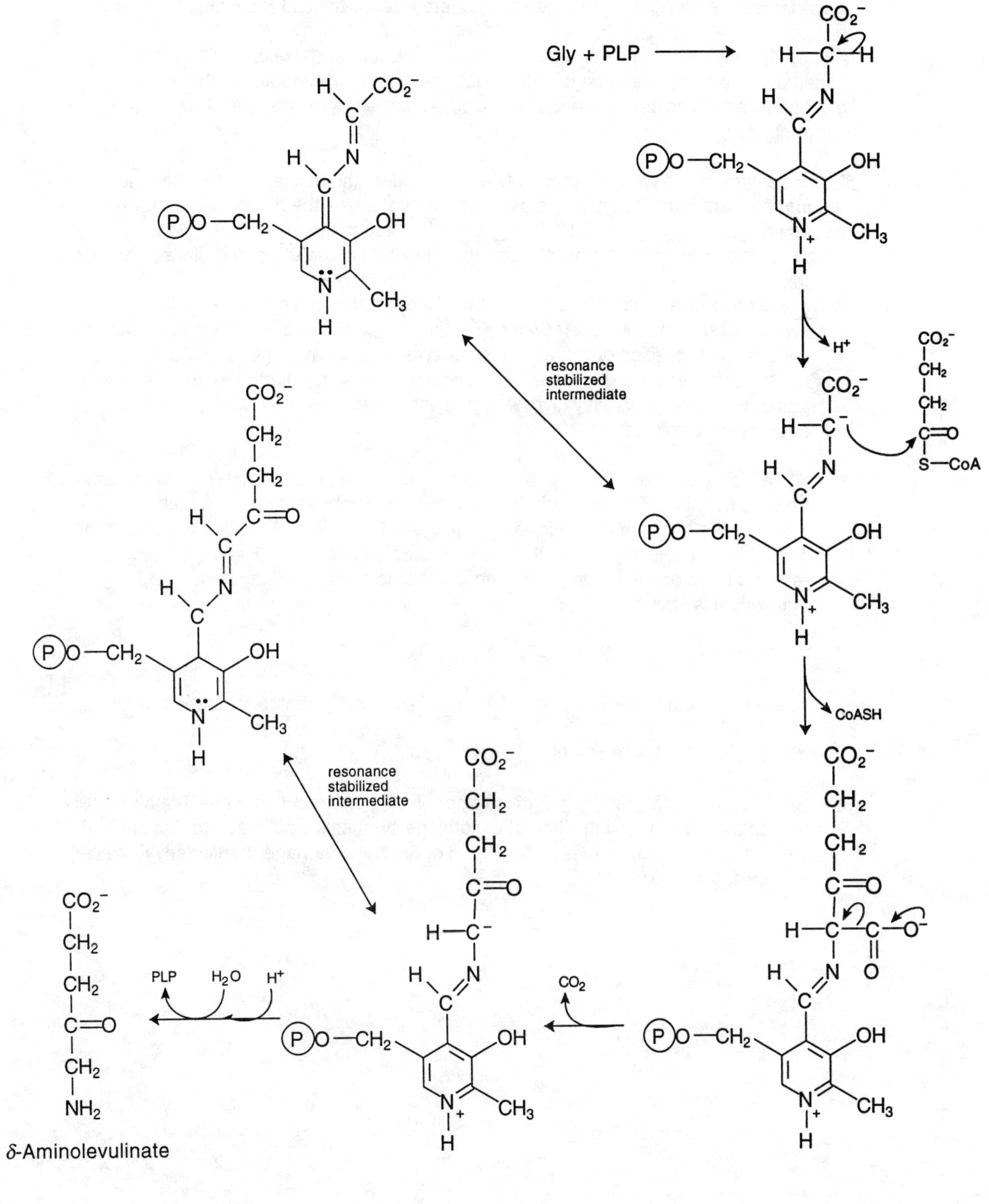
Gly + PLP
resonance stabilized intermediate
resonance stabilized intermediate
H+
CoASH
S—CoA
CO2
PLP
H2O
H+
δ-Aminolevulinate

20. The untreated diabetic animal synthesizes glucose primarily by hepatic gluconeogenesis utilizing amino acids derived from protein catabolism as the carbon source. Recall that the ATP required for gluconeogenesis is generated by fatty acid β-oxidation.

The amino acids are transaminated with α-ketoglutarate and oxaloacetate, forming glutamate and aspartate to be used in ureogenesis. The α-keto acids (those from glycogenic amino acids) then enter gluconeogenesis as pyruvate or as TCA cycle intermediates.

The urea cycle activity must increase to accommodate the increased flux of amino groups removed form the amino acids. Arginase catalyzes the hydrolysis of arginine yielding urea plus ornithine and is the rate-limiting step in the urea cycle. Hence you would predict an increased arginase activity in the liver of the untreated diabetic animal.

22. **a.** *L*-Glutathione is synthesized in successful steps catalyzed by γ-glutamylcysteine synthase and glutathione synthase. The information for the synthesis is dictated by the specificity of each enzyme. These enzymes are primary gene products (proteins), whereas *L*-glutathione is a secondary gene product. The formtaion of γ-glutamylglycine or cysteinylglycine is prevented because of the lack of the appropriate enzymes.

b. Glutathione, a substrate for glutathione peroxidase, is used catalytically in the cell as an antioxidant but is also a substrate consumed stoichiometrically by glutathione transferases. Loss of glutathione by conjugation with certain xenobiotics slated for excreation could compromise the cell's antioxidant defenses. Hence the cell requires an adequate supply of glutathione. One role of glutathione as an antioxidant is shown

$$2\ GSH + H_2O_2 \rightarrow G\text{--}S\text{--}S\text{--}G + 2\ H_2O \quad (Rxn\ 1)$$

where GSH and G–S–S–G are reduced and oxidized glutathione, respectively.

$$G\text{--}S\text{--}S\text{--}G + NADPH + H^+ \rightarrow 2\ GSH + NADP^+ \quad (Rxn\ 2)$$

Reaction 1 is catalyzed by glutathione peroxidase and reaction 2 by glutathione reducatase. Thus, an inhibition of glutathione synthesis would lead to depletion of cellular GSH and an increased likelihood of oxidative damage, particularly if the cell was oxidatively stressed.

26 Nucleotides

Chapter 26 Answers

2. A lone pair on an oxygen on Pi approaches the formyl carbon to produce tetrahydrofolate and the mixed anhydride of phosphoric and formic acids. The ADP approaches the phosphorus of the mixed anhydride to produce ATP and formate.

4. The decarboxylation of amino acids with the use of pyridoxal phosphate requires a free amino group on the amino acid. The amino group of the cysteine moiety is involved in an amide and cannot form the Schiff base intermediate with pyridoxal phosphate. Therefore, the decarboxylation of cysteine must occur by another mechanism.

6.

	Expected Results for Process:	
Labeled ATP	A	B
alpha 32p	Nonradioactive PRPP and Radioactive AMP	Nonradioactive PRPP and Radioactive ADP
gamma 32p	Radioactive PRPP (which loses radioactive Pi to form nonradioactive cyclophosphate on exposure to base, see prob. 26.5) and Nonradioactive AMP	Radioactive PRPP (when treated with base, see prob. 23.3, produces radioactive Pi and radioactive cyclophosphate) and Nonradioactive ADP

8. (1) The pathway for IMP starts with PRPP and the purine ring system is assembled on the PRPP, the PRPP is added after pyrimidine ring system assembly. (2) The purine ring system is assembled "atom by atom" from several sources in a "long" pathway while the pyrimidine ring is assembled from two components in a relatively "short" pathway.

10. Glycyl phosphate ($^{+}H_3NCG_2COPO_3^{2-}$) is the anticipated intermediate.

12. The ATP donates a phosphoryl (PO_3) group to the aldehyde oxygen. A phosphate is eliminated, taking with it what was the aldehyde oxygen.

14. Patients with Lesch-Nyhan syndrome have a deficiency in hypoxanthine-guanine phosphoribosyl transferase, an important enzyme in the salvage pathway for purines. When this enzyme is missing, the levels of phosphoribosylpyrophosphate (PRPP) become elevated and stimulate the synthesis of purines. These excess purines are degraded to uric acid, leading to severe gout. The gout can be treated with drugs like allopurinol, but this treatment has no effect on the severe mental retardation seen in these patients. The brain may lack a *ade novo* pruine biosynthetic pathway; it depends on the salvage pathway to produce purines, which are necessary for DNA replication. When a human is born, the rapid brain growth and development that normally occurs is inhibited if adequate levels of purine nucleotides are not produced. This genetic defect may eventually be treated with gene therapy.

16. Anticance drugs like the antifolates have greater toxicity toward rapidly growing cells because of the cells' demands for DNA replication. For example, methotrexate is a potent inhibitor of dihydrofolate reductase. The lack of synthesis of tetrahydrofolate prevents the formation of the various one-carbon tetrahydrofolate compounds. This limited one-carbon pool inhibits purine biosynthesis and, more importantly, the formation of methylene-THF, which is used in thymidylate synthase. The lack of dTTP synthesis in rapidly growing cells leads to death from lack of thymine. However, not all normal cells in human adults are nondividing. Some are growing rapidly, leading to the side effects observed in cancer patients treated with these compounds. Stem cells in bone marrow, epithelial cells of the intestine, and hair follicle cells are sensitive to the toxic effects of many anticancer drugs. This also explains why chemotherapeutic drugs are more toxic to children, especially when they are in a growth spurt.

18. The mechanism of the pyrophosphokinase is shown here.

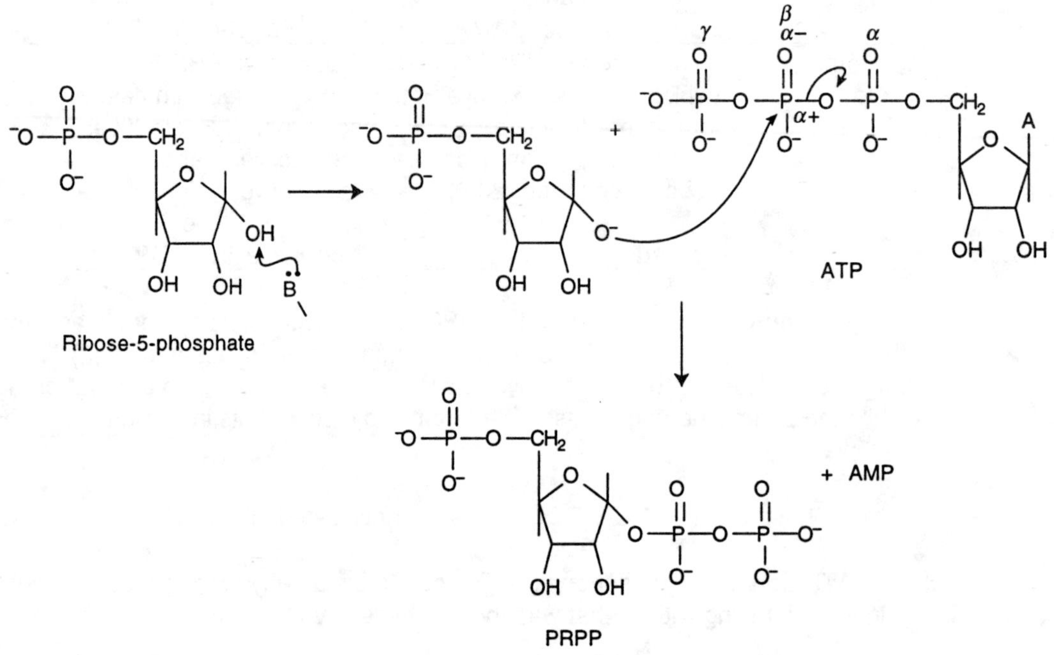

18. **(continued)**

If the γ phosphate of ATP is labeled with ^{32}P, then the phosphate group labeled in the illustration would be labeled in PRPP. When the radioactive PRPP$^\bullet$ is incubated with base, radioactive inorganic phosphate would be released (PRPP$^\bullet \rightarrow$ PRP + P$^\bullet$) and the phosphoribosylphosphate (PRP) would not be radioactive. If the β phosphate was labeled in ^{32}P-ATP, then the PRPP formed would be labeled in the phosphate attached to the 1'-carbon (PRP$^\bullet$P), and when PRPP was treated with base the radioactive phosphate would remain attached to the phosphoribosylphosphate (PRP$^\bullet$P $\rightarrow$ PRP$^\bullet$ + P). If the pyrophosphate group in PRPP was attached in the β configuration, then it would not be close enough for the deprotonated 2'-OH to carry out a nucleophilic attack on the phosphorous group, because the 2'-OH group and the 1'-PP group would be *trans* (see the mechanism in problem 5). Therefore, although pyrophosphate groups are very common in biochemical systems, the transfer of an intact pyrophosphate group such as seen with PRPP synthetase (pyrophosphokinase) is rare.

20. The chemical reaction mechanism for GAR synthase is

27 Integration of Metabolism in Vertebrates

Chapter 27 Answers

2. The term cycle in the phosphatidylinositol cycle is the transient nature of the existence of PIP_2, i.e., its breakdown and resynthesis. Any single or combination of components (phosphate, inositol, diglyceride, fatty acids, or glycerol) could be envisioned as completing the PIP_2 to breakdown products to resynthesis of PIP_2 sequence.

4. Using the binomial expansion and a 0.2:).8 M:H ratio:

M_4:	M_3H:	M_2H_2:	MH_3:	H_4
$(0.2)^4$	$(4)(0.2)^3(0.8)$	$6(0.2)^2(0.8)^2$	$4(0.2)(0.8)^3$	$(0.8)^4$
0.0016	0.0256	0.1536	0.4096	0.4096

6. Amino acid decarboxylases are expected to utilize pyridoxal phosphate as a cofactor.

8. Pathways are designed so that the equilibrium is very favorable for the conversion. The same pathway cannot be used for conversion in either direction because if it were thermodynamically favorable in one direction, it would be unfavorable by the same amount in the opposite direction. Also, the simultaneous occurrence of conversion in both directions would serve no purpose, as it would result in a futile cycle and waste energy. Regulatory enzymes are designed so that pathways never operate simultaneously in both directions.

   Fatty acid synthesis and degradation demonstrate these principles. In a liver cell, fatty acid synthesis takes place in the cytsol using acetyl-CoA carboxylase and a large, multifunctional polypeptide fatty acid synthase. The first committed step is acetyl-CoA caboxylase, which is highly regulated by hormonal control; this results in the phosphorylation of the enzyme. Citrate is transported from the mitochondria and is used to generate acetyl-CoA and reducing power in the form of NADPH (NADPH is used in biosynthetic reactions instead of NADH). Citrate activates the carboxylase while the end product, palmitate, inhibits the reaction. Fatty acid degradation occurs in the matrix of the mitochondria. A key point of regulation is on the uptake of the fatty acid into the matrix of the mitochondria. Malonyl-CoA, the product of the acetyl-CoA carboxylase, inhibits uptake and prevents the newly made palmitic acid from being degraded.

8. **(continued)**
    Gluconeogenesis utilizes many of the glycolytic enzymes, yet three of these enzymes in glycolysis have large negative free energy changes in the direction of pyruvate formation. These reactions must be replaced in gluconeogenesis to make glucose formation thermodynamically favorable. This allows both glycolysis and gluconeogenesis to be thermodynamically favorable and at the same time permits the pathways to be independently regulated so that a futile cycle does not exist. The first step in gluconeogenesis involves the movement of pyruvate into the matrix of the mitochondria, where it is converted to phosphenolpyruvate (PEP). PEP is transported back into the cytosol where it is converted by the glycolytic enzymes back to fructose-1,6-bisphosphate. Then two additional enzymes unique to gluconeogenesis allow glucose to be made. One of the most important allosteric effectors is fructose-2,6-bisphosphate, which activates phosphofructokinase, stimulating glycolysis, while at the same time inhibiting fructose bisphosphatase, inhibiting gluconeogensis. The levels of fructose-2.6-bisphosphate are under hormonal regulation in response to blood glucose levels.

10. A hormone receptor must do two things if it is to function properly:

    a. Distinguish the hormone from all other surroundings chemical signals and bind it with a very high affinity (K_d ranges from 1×10^{-7} M to 1×10^{-12} M).

    b. Upon binding the hormone, undergo a conformational change into an active form that can then interact with other molecules that initiate the molecular events leading to the hormone's elicited response.

    Proteins are the only macromolecules that can exhibit this kind of behavior (specific binding and conformation change).

    Acceptors are macromolecules that react to a receptor's conformational change by mediating enzyme activation (or inactivation). This is done by specific phosphorylation of proteins or production of regulatory molecules, both enzymatic events. Therefore, acceptors must be proteins also. The only exception to this (in a loose sense) is a DNA sequence to which a hormone-receptor complex binds and whose perturbation affects a distant promoter and thereby alters the transcriptional activity of that gene.

12. The precursor-product relationship serves the cause of hormone function is a variety of contexts, some of which are listed here.

    a. A polypeptide signal sequence must be present if the protein is to be transported into the endoplasmic reticulum and subsequently secreted.

    b. Additional polypeptide sequences are necessary for proper peptide chain folding (e.g., C peptide of insulin).

    c. Cleavage allows a control of hormones from inactive to active form (e.g., thyroxine).

    d. Production of a number of different hormones from the same precursor allows coordinate production of several hormones. Specific cleavage by the cell allows control of which peptides are produced (e.g., cleavage of prepro-opiocortin to corticotropin, β-lipotropin, γ-lipotropin, α-MSH, β-MSH, γ-MSH, endorphin, and enkephalin).

    e. A large precursor of the hormone can serve as a storage form (e.g., thyroglobulin).

14. The same amino acid, named as 5-oxoproline, is an intermediate in the γ-glutamyl cycle (figure 25.15). The mane pyroglutamate suggests formation involving dehydration via heat from glutamate.

16. Vitamin D can be considered both a hormone *and* a vitamin. Its mode of action is like many other steroid hormones (forming a receptor-hormone complex that activates transcription of specific genes in the nucleus), and it is synthesized in the body where it acts at a distant location. Vitamin D_3 is formed in the skin of animals through the action of ultraviolet light on 7-dehydrocholesterol. Vitamin D can also be taken in the diet (commonly as a vitamin supplement in milk) and is then considered a vitamin.

28 Neurotransmission

Chapter 28 Answers

2. If there was no acetylcholine esterase there would be a constant concentration of acetylcholine in equilibrium with acetylcholine bound to receptors in the postsynaptic membrane. A change (increase) in acetylcholine is required to generate a new action potential in the recipient cell. Sequential signal transmission through the same synapse requires the depletion of the chemical neurotransmitter.

4. All of the non-peptide neurotransmitters are amines or amino acids (one is a sulfonic acid). The majority of the confirmed neurotransmitters are amines. All of the amines listed are confirmed neurotransmitters.

6. Monoamine oxidase is a flavoprotein. An initial intermediate is the production of an imine with the reduction of an enzyme bound FAD. To complete the reaction the imine is hydrolized with water (the aldehyde oxygen originating with the water) and the $FADH_2$ is oxidized with molecular oxygen to produce hydrogen peroxide.

8. The fate of the two oxygen atoms in O_2 is different. With nonoamine oxidase both of the O_2 atoms end up in the same product (H_2O_2) while with dopamine hydroxylase the two oxygens end up in different products (water and norepinephrine).

10. The most obvious difference between the mentioned ions is their relative sizes. As you proceed down a group in the periodic chart the ions are larger. The simplest explanation involves the pore "gauging" the diameter of the hydrated ion to be transported. Alternatively, rather than just considering the charge and size the concept of charge density on the ion's surface may be recognized by the pore.

12. These criteria include but are not limited to

    (1) The ability of exogenously added putative transmitter to excite the postsynaptic cell in the same manner as occurs when the presynaptic cell is stimulated electrically.

    (2) The presence of high concentrations of the resumed transmitter in the presynaptic nerve terminal.

    (3) The release of the compound from the nerve terminal upon stimulation of the presynaptic cell.

    (4) The presence of specific receptors for the compound in the postsynaptic membrane.

14. The ball-and-chain model or "stopper-on-a-string" mechanism can explain most of the experimental data on the K$^+$ channel. Previous experiments had shown that the N terminal is involved in inactivation, so numerous mutations in this region were tested in the *Xenopus* oocyte expression system. The first 20 amino acid residues form a "ball" with a hydrophobic core and positively charged amino acids on the surface. These positively charged residues probably help bind the ball to the inside surface of the pore by interacting with the negatively charged amino acids near the opening. The next 20 amino acids act as a "chain" that can swing the ball in or out of the pore opening. Even the free ball can inactivate a mutant that lacks the ball if high concentrations are added. This model (figure 28.12) will allow scientists to design experiments to determine the details of this mechanism. Similar approaches to other ion channels will determine whether this is a common mechanism for all these pores.

16.

When acetylcholinesterase is inhibited by compounds like parathion, the levels of acetylcholine in the synaptic cleft remain high and do not allow the depolarizing signal to be switched off. The acetylcholine has to be hydrolyzed to allow the membrane to become repolarized to allow the nerve to continue working.

29 Vision

Chapter 29 Answers

2. The hexasaccharides and palmitoyl groups are probably involved in keeping the appropriate portions of the peptide associated with the disk lumen or the lumen membrane, respectively.

4. Compounds that absorb visible light and are used in light-detection systems have many alternating double bonds (conjugated double bonds), such as retinal, or even as in chlorophyll found in plants. This allows visible light to be absorbed and the compound to become excited to a higher energy state where it can undergo oxidation (chlorophyll) or isomerization (retinal).

6.

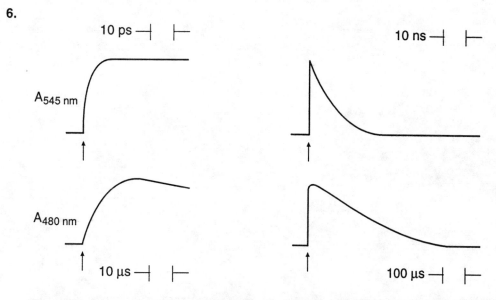

The absorbance changes at 545 nm are due mainly to the formation and decay of bathorhodopsin; those at 480 nm, mainly to metahodopsin I. (Vertical arrows indicated flashes.)

8. The Schiff's base linkage between the retinal group and opsin is protonated in metarhodopsin I but not in metarhodospin II.

10. When retinal binds to the rhodopsin, if forms a Schiff's base with a lysine residue. The nitrogen atom in the Schiff's base linkage is protonated and positively charged. When the absorbs light, the electron density of the nitrogen increases and the positive charge moves to the opposite end of the molecule (see figure 29.8 in the text). This redistribution of charge makes the absorption properties very sensitive to charged groups nearby. There is a glutamic acid residue that forms an anionic counterion with the protonated nitrogen in the Schiff's base. If this glutamate is changed by site-directed mutagenesis, the pK value of the base is changed and the Schiff's base in the mutant rhodopsin is not protonated. This lack of protonation changes the absorption maximum from 500 nm to 380 nm, the absorption maximum seen with free retinal in solution.

30 Structures of Nucleic Acids and Nucleoproteins

Chapter 30 Answers

2. Recall from organic chemistry that an ester results from the condensation of a carboxylic acid with an alcohol. A nucleotide is already a phosphomonoester, arising from the condensation of an acid (phosphoric acid) with a sugar alcohol (ribose). A phosphodiester linkage, then, results from the condensation of an acid (phosphoric acid) with two alcohols (the riboses) to form a dinucleotide.

4. The enol form of thymine is shown below.

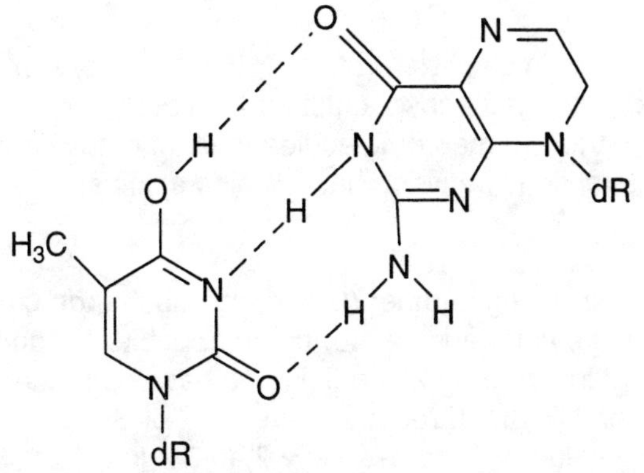

The enol form of thymine can hydrogen bond with guanine to form a base-pair similar in overall dimensions to a GC base-pair, as shown above.

6. **a.** Note the designation of major and minor grooves in the text. The following hydrogen bond donors and acceptors are available in the major groove of B-form DNA:

Adenine: N^7; N^6
Cytosine: N^4
Guanine: N^7, O^6
Thymine: O^4

The following hydrogen bond donors and acceptors are available in the minor groove of B-form DNA:

Adenine: N^3
Cytosine: O^2
Guanine: N^3; N^2
Thymine: O^2

(Note also the possibility of van der Waals interactions in the major groove with C^5 and C^6 of the pyrimidines and C^8 of purines.)

b. Although the number of electronegative atoms capable of forming hydrogen bonds in the major and minor groove are similar, access to those in the minor groove is hindered by the ribose moiety (see Figure 30.x). Because hydrogen bond donors and acceptors are more accessible in the major groove, most (but not all) proteins interact with this face of the DNA double helix.

c. Positively charged residues on the protein stabilize protein-DNA interactions in two ways: by neutralizing the negatively charged phosphodiester backbone, thus permitting intimate contact between the two macromolecules, and, in many cases, by forming specific electrostatic interactions between basic amino acids and phosphate groups.

8. **a.** Since the backbone is not broken, the linking number (L) remains unchanged, and any change in supercoiling (S) must be compensated for by an opposite change in twist (T). The DNA molecule at right in figure 30.20 contains 40 base-pairs that are disrupted. When two negative supercoils are introduced, the value of S changes from 0 to -2, and, consequently, the value of T increases by 2, to 34. In B-form DNA (10 bp per turn), the additional 2 twists permit 20 base-pairs to re-form. Therefore, in the 360 bp DNA illustrated, 20 base-pairs (of the original 40) would remain disrupted.

b. Z-DNA is left-handed instead of right-handed. Therefore, the twist (T) introduced by a turn of Z-DNA turns in the opposite direction compared to B-DNA, and Z-DNA has the effect of decreasing the twist (T) by one for every 12 bp of Z-DNA. When the linking number is constant (*i.e.* the backbone is not broken) formation of 12bp of Z-DNA will permit an additional 10 bp of B-DNA to form. In the molecule illustrated in figure 30.20, assuming S is unchanged, only 18 bp of single-stranded DNA would remain (40 bp originally - 12 bp Z-DNA - 10 bp B-DNA).

10. At high pH, guanine and thymine would be deprotonated and have a negative charge, while at low pH, adenine, guanine and cytosine bases would be protonated and have a positive charge. In addition to impeding the normal Watson-Crick hydrogen-bonding between bases, these charges in the hydrophobic interior of the duplex would disrupt its structure due to charge-charge repulsion between the two strands. At neutral pH, the bases would not have any charge and the duplex would have its greatest stability. Living systems maintain their pH around 7, which preserves the duplex structures found in their nucleic acids (DNA and RNA).

12. The finding that scrapie was caused by a proteinaceous particle was unexpected as it was believed that only nucleic acids could be precisely replicated.

    To demonstrate that scrapie is caused by protein(s), one would need to demonstrate that the infectivity of the scrapie particle is unaffected by procedures which modify nucleic acids, and that it is abolished by procedures which affect only protein. For example, the scrapie agent is relatively insensitive to ultraviolet light and nuclease digestion. Scrapie particle infectivity is sensitive to protease digestion, high temperature inactivation and phenol extraction.

14. The fact that the single-strand specific S1 nuclease occasionally cleaves AT-rich regions of double-stranded DNA reveals that these regions are not fully double-stranded at all times, and that the structure of double-stranded DNA is not static. The rigid, linear model illustrated in many textbooks is actually one of many conformations. Depending on its base composition, DNA can bend and curve, and the base-pairs may temporarily pull apart and then reform, in a process known as "breathing". AT-rich sequences are particularly prone to transient denaturation, because of the weaker nature of AT base-pairs. S1 nuclease recognizes and cleaves such regions when they briefly acquire single-stranded character.

16. This equation incorporates a number of the variables which we know affect the stability of the double helix. For example, cations stabilize the double helix by shielding the anionic phosphodiester backbone. Thus, we see that T_m increases with the log of M, the concentration of monovalent cations. GC base-pairs contribute more to the stability of the DNA duplex than AT base-pairs, represented in this equation by a term incorporating %G+C. Formamide tends to destabilize double-stranded nucleic acid structures because it disrupts the hydrophobic forces holding together the bases, and provides plenty of hydrogen bond forming groups. Therefore, T_m decreases with increasing formamide concentrations. Finally, the length, L, of a DNA also affects its melting temperature, since a longer strand would have significantly more hydrogen bonding and stacking interactions to overcome before it could melt. Of course, these are only a few of the many factors governing DNA duplex formation, but the ones in this equation are those which are most frequently varied experimentally.

18. Highly repetitive DNA becomes double-stranded at a lower C_0t value than the moderately repetitive or unique sequence fractions. Thus by allowing a dilute (low C_0) sample of DNA to renature for a short time (small t) and then subjecting it to chromatography on hydroxyapatite, you should be able to separate a double-stranded, highly repetitive fraction from the rest of the DNA.

20. The regulation of gene expression in eukaryotes is not very well understood. However, the arrangement of genes in eukaryotic genomes, specifically the clustering of functionally related genes, could sequester transcription factors and other accessory proteins that help to regulate the expression of these genes.

31   DNA Replication, Repair, and Recombination

Chapter 31 Answers

2.   In this experiment, labeling of any cellular component other than DNA would be undesirable because it would complicate the interpretation of the experiment. Thymidine is incorporated only into DNA (unlike adenosine, guanosine or cytidine, all of which would also be incorporated into RNA). The pathway for incorporation of [^{3}H]-thymidine into TTP, which is then used for DNA synthesis, involves the following reactions (review in Chapter 28):

Thymidine + ATP $\rightarrow$ DTMP + ADP   (Thymidine kinase)

dTMP + 2 ATP $\rightarrow \rightarrow$ dTTP + 2 ADP   (Nucleoside monophosphate and diphosphate kinases)

Thymidine was used in these experiments rather than DTTP because nucleotides, with their charged phosphates, usually cannot cross cell membranes. Thymine, the free pyrimidine base, might have worked as well, since bacteria possess a thymine pyrophosphorylase enzyme which catalyzes the formation of thymidine from thymine and deoxyribose-1-phosphate.

^{3}H was used as the radioactive label in these experiments because the short distance traveled by its low-energy β particle would keep the spot of exposed film to a minimum size and make the microautoradiograph easier to interpret. Both ^{3}H and ^{14}C emit β particles, but that emitted by ^{3}H has a lower energy than the ^{14}C β particle. Although ^{3}H is more difficult to detect (longer exposure times are needed), the particle does not travel as far as it would for ^{14}C. The high-energy β particle emitted by ^{14}C can travel a considerable distance before hitting the film, obscuring the details of the replication eye visualized in Figure 31.5.

4. **a.** The location of the origin of replication must be closest to the gene with the highest average number of copies per cell, in this case gene *c*.

    b. Taking into account the plot of gene frequency and the physical map of the chromosome described, we can see that the average number of copies/cell of genes on both sides of the origin (gene *c*) declines gradually from 2 to 1. This pattern would be expected if replication were bidirectional. If replication were instead unidirectional, the pattern of decline in gene frequency (assuming the same genetic map) would be:

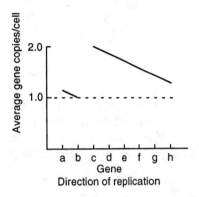

6. Short chains of DNA are inherently more difficult to synthesize accurately than long chains because of the absence of long-range stacking interactions. Hydrogen bonds are weak, and a few base-pairs of DNA cannot form a very stable double helix even when the bases are correctly paired. As more bases are added, the double helical structure is stabilized by the increase in stacking interactions. (Another way of saying this is that base-pairing is a cooperative process.) Because the first few bases are inherently unstable, their stability cannot be verified by the polymerase. The use of a primer increases the fidelity of DNA synthesis by providing a somewhat more extensive stacked double helix to which the first few bases are added, allowing the proofreading exonuclease activity of DNA polymerase to more accurately evaluate the stability of the newly synthesized DNA.

    Because the primer itself cannot be checked as it is being synthesized (for the reason mentioned above), it is likely to contain some errors and must be subsequently repaired. It may be more efficient to replace the entire primer than to carefully repair only the mismatches in the primer-containing region. The reason why the primer is usually RNA and not DNA may be because RNA-containing double helices are readily distinguished from DNA-DNA duplexes (recall that RNA-DNA hybrids adopt an A-form double helix). Recognition of the RNA-containing strand because of its altered structure facilitates the removal and replacement of the RNA primer.

    It is thought that, very early in evolution, RNA was used as the genetic material, and not DNA. The enzymes for making RNA primers might have been present when the first DNA polymerases evolved and were natural candidates for primer synthesis.

8. A DNA polymerase which catalyzes hyperaccurate replication might be evolutionary disadvantageous for the following reasons: First, it would function more slowly than the wild-type DNA polymerase (additional time would be required for proofreading as well as excising and replacing misincorporated nucleotides), allowing cells with a slightly less accurate polymerase to outcompete the hyperaccurate mutant. Secondly, some misincorporation is beneficial as it allows the organism to adapt to a changing environment by generating genetic diversity from which advantageous mutations may arise. Finally, hyperaccurate replication is energetically expensive, since it involves the hydrolysis of additional nucleotides, in many cases unnecessarily.

10. The small amount of DNA polymerase activity was not sufficient to account for the mutant's normal rate of replication. This suggests that another polymerase (DNA Pol III) must be the primary enzyme in replication. However, the Pol I mutant was deficient in repair synthesis (could not repair UV damage, *etc.*), suggesting that this must be one of the major roles of DNA polymerase I. Cairns and DeLucia were lucky because a mutant completely lacking DNA polymerase I activity could not survive (a lethal mutation). In addition to its role in DNA repair, Pol I is important for the maturation of Okazaki fragments, namely removal of the RNA primer and "nick translation".

12. The *dnaB* helicase reaction results in unwinding two strands of DNA. Disrupting the corresponding hydrogen bonds requires energy, supplied by ATP. *dnaB* protein does not alter the linking number of the DNA, as it does not break the phosphodiester backbone. Rather, it uses the torsional stress caused by the negative supercoiling introduced by DNA gyrase to facilitate the unwinding of DNA at the replication fork. (T increases while W decreases, keeping L constant.)

14. Several steps in the initiation of DNA replication in *E. coli* require the hydrolysis of ATP. ATP hydrolysis in the initiation steps is largely required for the dramatic conformational changes in the DNA at the origin, especially strand separation. ATP hydrolysis in some steps of initiation also drives the specific assembly of some of the multiprotein complexes required for replication. Primer synthesis requires energy, supplied by the hydrolysis of the NTPs incorporated in the primer.

    The elongation reaction requires energy, for replication fork movement and for the formation of the energy-rich phosphodiester bonds. The energy for replication fork unwinding is supplied by ATP hydrolysis. However, the energy for nucleotide addition is supplied by hydrolysis of the dNTPs, releasing pyrophosphate. The hydrolysis of each dNTP drives its incorporation into the nascent strand.

    DNA ligase requires energy to form a phosphodiester bond between a 3'-OH and a 5'-phosphate while sealing nicks in the lagging strand. While some DNA ligases use ATP for this purpose, *E. coli* DNA ligase uses NAD^+.

    In addition to the direct energy requirements listed above, there is also an ATP requirement for DNA gyrase to introduce and maintain the correct degree of negative supercoiling for DNA synthesis to proceed.

16. UV light causes the formation of pyrimidine dimers, which are remove by a general excision repair mechanism. Single-strand breaks occur as the first step in this repair process. The damaged DNA is removed and replaced, and the nicks are ligated. A short time after exposure to UV light, many single-strand breaks are observed in the DNA (generating the small fragments of single-stranded DNA on the alkaline sucrose gradient, graph (a)). After many hours the DNA is repaired and only large segments of single-stranded DNA are found.

    The data from the patient with xeroderma pigmentosum would resemble those in graph (b), as this patient is deficient in excision repair. No single-strand breaks appear in the DNA and the single-stranded DNA remains of high molecular weight. The inability to remove pyrimidine dimers leads to the early development of skin cancers and eventual blindness and death.

18. Some of the sister chromatids in Figure 31.3b clearly exhibit a mosaic pattern of radioactive grains (*e.g.* that furthest to the right), suggesting that each chromosome is not uniformly labeled throughout its length. Furthermore, the pattern of labeling in one of the sister chromatids appears to be the inverse of the labeling in the other. This reversal of labeling pattern is due to the exchange of genetic information between the sister chromatids.

20. Retroviruses (like the AIDS virus, HIV) first convert their RNA genome into DNA using the unique enzyme reverse transcriptase (an RNA-dependent DNA polymerase which makes a DNA copy of an RNA molecule)and using a cellular tRNA as a primer. This DNA is made into a double-stranded circular DNA which enters the nucleus and integrates into the host genomic DNA, where it can remain latent for a long time (called a provirus). The integrated viral sequence is similar to any other eukaryotic gene. When the provirus is activated, it produces an infectious mRNA that is packaged with viral proteins encoded by the mRNA, including reverse transcriptase. The virus particle is released from the cell surface by a budding process, and goes on to infect other cells.

22. The minimal sequences required for an artificial chromosome to replicate in yeast would be a centrosome, an origin of DNA replication and telomeres at either end. The centrosome would be required for correct segregation of the artificial chromosome upon cell division. The need for an origin of replication is obvious. Telomeres would be required to prevent catastrophic shortening at the ends of the artificial chromosome.

32   DNA Manipulation and Its Applications

Chapter 32 Answers

2.   Four separate reaction tubes would be prepared, each of which contains the RNA preparation to be sequenced, a specific primer, reverse transcriptase, dNTPs and a small amount of one of the dideoxy-NTPs. The products of the reverse transcriptase reaction would be labeled by including a radioactive dNTP or by labeling the primer used in the reaction. After incubation of the reverse transcriptase reaction, the reactions would be stopped and the products analyzed by denaturing polyacrylamide gel electrophoresis. The results would be identical in appearance to a DNA sequencing gel. The RNA that was sequenced would be expected to have the sequence complementary to the DNA sequence deduced from the gel.

4.   The PCR can be carried out with shorter primers, as long as the annealing temperature is less than the predicted melting temperature for that primer-template combination. The problem with shorter primers, especially when performing the PCR with genomic DNA, is that there is an increased probability of nonspecific hybridization of the primer, potentially leading to PCR artifacts. Other considerations which go into PCR primer design are the G+C content of the primer, degeneracy of the primer, and complementary within and between primers. It is important that primers used in the same PCR reaction be similar in size and G+C content, to ensure that they anneal at similar temperatures. The "degeneracy" of the primer refers to primers that are designed by back-translation of an amino acid sequence into a population of different possible coding sequences. Obviously, the more degenerate the primer, the greater the probability that it might hybridize to a sequence other than the desired gene. Finally, complementary within and between PCR primers can lead to undesired side reactions and/or competition with the desired template.

6. The restriction map of this fragment is given below:

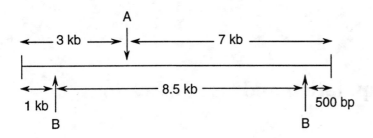

The data given in the problem tell you that the 10,000 bp fragment has one cleavage site for enzyme A, 3,000 bp from one end, and two cleavage sites for enzyme B. The only possible arrangement for the two cleavage sites for B that would give the indicated fragments in a double digest of A and B together is the arrangement shown.

8. DNA digested with *ClaI* and then labeled with the Klenow fragment of DNA polymerase I and [α-^{32}P]dCTP would have the structure shown below. This strategy could not be used for efficient radioactive labeling of fragments generated by *KpnI* because the 3'-overhang generated by *KpnI* cannot serve as a template for DNA synthesis.

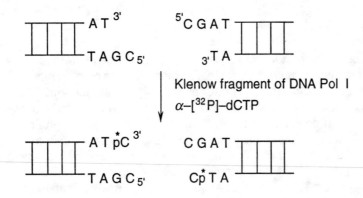

10. A transgenic mouse line carrying a mutant CFTR gene could be used to study the pathology and progression of cystic fibrosis without harming humans who suffer from this disease. One of the most beneficial applications of these mice would be to develop new treatments and to evaluate therapeutic strategies for cystic fibrosis.

12. To reduce the hybridization stringency, researchers choose conditions which stabilize double-stranded DNA, allowing sequences which share only partial complementary to form base-pairs. Examples of such conditions include reducing the hybridization temperature, or increasing the salt concentration in the hybridization solution. A reduction in temperature helps stabilize the double-stranded structures formed between heterologous DNAs, which have a lower melting temperature than perfectly paired DNA. The salt helps neutralize the repulsion between phosphodiester backbones, again facilitating the formation of a duplex between sequences of limited complementary.

14. Fetal DNA isolated from amniotic fluid would be digested with *MstII* and subjected to
     agarose gel electrophoresis. The Southern blotting procedure could be performed,
     using the normal β-globin gene as probe. The expected results are illustrated below,
     with normal DNA at left, DNA from a sickle cell patient on the right, and a carrier (with
     sickle cell trait) in the center.

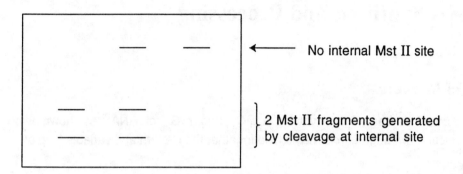

    ←——— No internal Mst II site

    2 Mst II fragments generated
    by cleavage at internal site

The Southern blot results illustrated above demonstrate that normal human DNA (left
lane) yields two *MstII* fragments that hybridize with the β-globin gene probe, while DNA
carrying the sickle cell mutation (right lane) yields only one larger *MstII* fragment,
because it has lost the *MstII* within the gene.

Problems with this procedure include potential difficulties in isolating sufficiently large
quantities of fetal DNA, and likely contamination of the fetal DNA with maternal DNA.
For example, if the mother had sickle cell trait, the baby was homozygous for the
disease, and a mixture of the two DNAs were digested with *MstII*, the results would be
difficult to distinguish from those expected of a sickle cell carrier, like the individual
whose DNA is shown in the center lane.

16. If you transformed *Saccharomyces cerevisiae* with a linear fragment of a yeast gene into
     the middle of which was inserted an antibiotic resistance gene, and then selected for
     resistance to that antibiotic, you could isolate a strain in which the original gene was
     nonfunctional. By observing the phenotype caused by the inactivation of this gene, you
     could find out information about the role of its gene product inside the cell. Such "knock-
     out" mutations are extremely useful for the study of gene function.

18. Telomeric DNA has very few or no recognition sites for most restriction enzymes. Thus,
     when chromosomal DNA is digested with a restriction enzyme, the telomeres would be
     found in fragments which have one end corresponding to a restriction site within the
     chromosome, and one end which represents the ragged end of the telomeric DNA. The
     telomeric end of this fragment would not have the correct overhang for annealing and
     ligation to the cloning vector, so it is likely that such fragments would be under-
     represented in a genomic library.

33   RNA Synthesis and Processing

Chapter 33 Answers

2. A structure for the base triplet involving G_{22}, C_{13} and m^7G_{46} of tRNA^{Phe} is shown below. This particular "*ménage à trois*" helps hold together the D-stem and variable loop of tRNA^{Phe}.

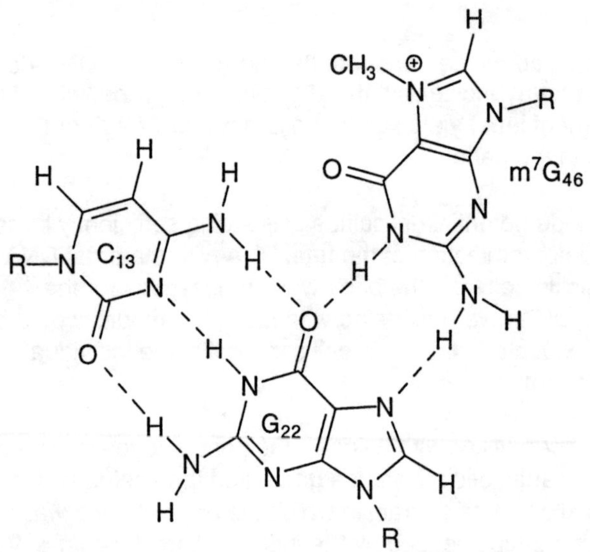

4. As in DNA replication, RNA polymerase does require topoisomerase activity to make unwinding of the DNA feasible. The main reason why helicases and single-strand binding proteins are not required in transcription is because RNA polymerase only unwinds a short stretch of DNA (10-20 bp). After the initial melting, as the RNA polymerase moves along its template, there is always one base-pair being formed (behind the transcription bubble) for each base-pair opened (ahead of the transcription bubble). In contrast, the replication apparatus must eventually separate the two strands along their entire length.

6. The –35 and –10 (Pribnow) sequences are underlined, and the likely initiating nucleotide is doubly underlined. The –35 box is a close match to the consensus, and it is unlikely that it can be improved through mutagenesis. The Pribnow box in this promoter, on the other hand, is a poor match to the consensus. Changing the second A to a T would better match the consensus, and should make this a stronger promoter.

GCGGGATCGTTGTATATTTC<u>TTGACA</u>CCTTTTCGGCATCGCCC<u>TAAAAT</u>TCGGC<u><u>G</u></u>TCCTCATATTG
                                                ↓
                                                T

8. mRNA in bacteria is degraded extremely rapidly, with a turnover rate of approximately 50% in 1 to 3 min, and thus does not accumulate. The 3% of the total RNA present at any time represents that mRNA which has just been synthesized but not yet degraded. However, rRNA and tRNA are considerably more stable and accumulate. Even though these RNAs together account for roughly 50% of the total RNA synthesis, they represent greater than 95% of the total RNA present at a given time.

10. The TATA-binding protein, in addition to flattening and widening the DNA by interacting (atypically) with the minor groove, also introduces a sharp bend ($>100°$) into its recognition sequence upon binding. This bend would bring the DNA upstream and downstream of the TATA box closer together, promoting interactions between proteins bound to upstream elements and those near the start site of transcription. In addition, the flattening and widening of the DNA double helix, in conjunction with the bending, may promote the strand separating activity of the transcription initiation complex.

12. T4 DNA ligase uses ATP to provide the activation energy for phosphodiester bond formation between fragments with 5'-phosphates and fragments with 3'-hydroxyls. In the ligase reaction, the positioning of the fragments to be joined is governed by hydrogen bonding and base-stacking interactions between strands of the substrate(s).

   In mRNA splicing, ATP powers the correct interaction and folding of the spliceosomal RNAs as well as the hnRNA. In the splicing reaction, the positioning of the nucleotides to be joined is controlled by this three-dimensional structure, which requires ATP for its formation. Here, the energy for phosphodiester bond formation is coupled with the hydrolysis of an energetically equivalent bond and does not require ATP.

14. a. The A at position 6 of the branch point TACTAAC sequence is the A at which the lariat forms its 2'-5' linkage. Changing this A to a C would prevent lariat formation, inhibiting the first step of mRNA splicing. Intron-containing RNA would accumulate.
   b. Changing the G at the 5'end of the intron to a T would also abolish splicing of this intron and intron-containing RNA would accumulate.
   c. Changing the G at the 3'-end of the intron to an A might affect spliceosome formation, but if the spliceosome can form, it would abolish the second step in mRNA splicing. The lariat intermediate would be expected to accumulate in this case, unless a "cryptic" splice acceptor site (meaning a sequence which is further downstream and is not normally used as a splice site, but bears some resemblance to the splice site consensus) could replace the mutated site.

16. There are three possible mechanisms for apolipoprotein-B mRNA editing based on similar changes which occur in DNA repair:

   a. Base modification: simply deaminating the specific cytosine residue would leave a uracil in its place. To test this mechanism, use ^{3}H-cytidine triphosphate for RNA synthesis, and show that the label is retained in uracil after RNA editing.

   b. Base excision: the cytosine N-glycosidic linkage could be specifically cleaved, leaving an abasic site in the mRNA. Uracil could then be incorporated, much like the base exchanges that occur in some tRNA modification mechanisms. To test this mechanism, one would first show that the cytosine ring is not retained, using ring-labeled CTP as described above. Using α-^{32}P-CTP for RNA synthesis, one would demonstrate that the label is retained after RNA editing, suggesting that the RNA backbone is not broken during processing.

   c. Patch repair: a guide RNA could be used as template to direct the incorporation of one or more bases after cleavage of the RNA on both sides of the cytosine to be replaced. To test this mechanism, the editing reaction could be shown to require NTPs and to result in the incorporation of α-^{32}P-NTPs during the course of the editing process.

18. The simplest explanation for these findings is that TBP is directly involved in transcription by all three RNA polymerases, despite the fact that TATA boxes have only been recognized in RNA polymerase II promoters.

34 Protein Synthesis, Targeting, and Turnover

Chapter 34 Answers

2. The anticodon sequence of this tRNA must be IGA. Note that the anticodon sequence, like the codons with which it pairs, is given in the 5' to 3' direction. The inosine residue (I) at the 5'-end of the anticodon can form wobble base-pairs with C, U or A at the 3'-end of the codons, as shown below:

tRNA Anticodon:	3'-AGI-5'	3'-AGI-5'	3'-AGI-5'
Codon:	5'-UCA-3'	5'-UCC-3'	5'-UCU-3'

4.
Leucine codon:	Possible tRNA anticodon(s):
UUA	UAA (AAU is an incorrect answer, although 3'-AAU-5' is acceptable because the correct direction of the chain is given.)
UUG	UAA, CAA
CUA	UAG, IAG
CUC	GAG, IAG
CUG	CAG, UAG
CUU	AAG, GAG, IAG

It is unlikely that a tRNA^{Leu} with the anticodon IAA exists, because IAA could recognize the phenylalanine codons UUU and UUC in addition to the leucine codon, UUA. Such a tRNA would promote misreading of phenylalanine codons by inserting leucine.
It is important to remember that anticodons and codons are always given in the 5' to 3' direction, but the interaction between the two is antiparallel, just like the two strands of DNA molecules. (As indicated above, the only time it is acceptable to give nucleic acid sequences 3' to 5' is if the direction is obvious, either because it is clearly complementary to another 5' to 3' strand, or because the 3' and 5' ends are labeled.)

6. There are three isoleucine codons in the genetic code: AUU, AUC, and AUA. Normally, we expect tRNA isoacceptors which read rare codons themselves to be rare. The tRNA^{Ile} isoacceptor which could read AUA codons should have the anticodon sequence IAU, and we would expect this tRNA to be underrepresented among *E. coli* tRNAs. (Note that a UAU anticodon sequence, which could also read methionine codons, would be surprising in tRNA^{Ile}.) On the other hand, assuming no unusual modifications of the anticodon sequence take place, the tRNA^{Ile} isoacceptor(s) with GAU and possibly AAU anticodons are likely to be abundant.

8. The energy for peptide bond formation is released upon the hydrolysis of the ester bond between the P site-bound amino acid and its tRNA. The ribosome-associated peptidyl transferase activity couples peptide bond formation to the hydrolysis of the energetically similar aminoacyl-tRNA bond (an ester). The energy for the formation of the aminoacyl-tRNA bond, in turn, came from the hydrolysis of ATP during the aminoacyl-tRNA synthase reaction.

10. Assuming a single base change for each step, the wild-type codons can in each case be deduced by simple inspection of table 34.x. For example, in part (c), the relationship Leu → Met allows only two possible identities for the wild-type leucine codon: UUG or CUG. A single point mutation of the first nucleotide of only these two leucine codons could give rise to a methionine codon (<u>A</u>UG). In the same manner, the relationship Leu → Ser is consistent with two Leu codons, UUA or UUG (giving rise, through a mutation at the second nucleotide, to U<u>C</u>A/G serine codons). The overlap between these conditions unambiguously identifies the original Leu codon as UUG. (The Leu → Val mutation is not informative, since any of the six leucine codons can be mutated to a valine codon by a single nucleotide change.)

    Similar treatment of other relationships provides the following codon identifications:
    a. Gln (CAG) → Arg (C<u>G</u>G) → Trp (<u>U</u>GG)
    b. Glu (GAA) → Lys (<u>A</u>AA) → Ile (A<u>U</u>A)
    c.

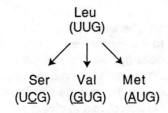

    d.

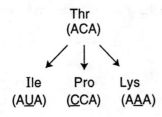

12. The structure of phenylalanine is very similar to that of tyrosine, except that it lacks the phenolic hydroxyl group. However, phenylalanine is activated 1.5×10^5 times less efficiently than tyrosine by tyrosyl-tRNA synthetase. The hydroxyl group of tyrosine participates in two hydrogen bonds, to amino acids Asp 176 and Tyr 34 of tyrosyl-tRNA synthetase. Formation of these hydrogen bonds is critical for the synthetase's ability to discriminate between tyrosine and phenylalanine.

14. Fusidic acid sensitivity is dominant because of the fact that each molecule of mRNA is translated by many ribosomes at one time. If one ribosome in a polysome is frozen in place by the failure of a molecule of fusidic acid-sensitive EF-G to cycle off of the ribosome, all the trailing ribosomes will be unable to continue forward because of the stalled ribosome in their path. As a result all protein synthesis soon stops and the cells die.

16. If one of the two promoters (P_2 in the figure below) is downstream from the AUG at the beginning of the transit peptide, mRNA transcribed using this promoter would initiate translation at the closest AUG (which may or may not be within the transit peptide coding region, as shown below). This mRNA would encode a polypeptide which lacks a transit peptide, and consequently could not possibly be imported into mitochondria ("cytoplasmic peptide"). Translation of the longer mRNA originating from P_1 would yield a protein with its transit peptide, which could be imported into mitochondria.

    (See page 104 for figure.)

18. PEST sequences are regions rich in proline, serine, and acidic residues that are common in short-lived eukaryotic proteins. They may be recognized directly by the protein degradation machinery, or they may promote unfolding of the proteins in which they are contained, indirectly leading to degradation. Regulatory proteins tend to be rather short-lived, due to the nature of their function (the cell must be able to shift to a new pattern of gene expression as conditions warrant). Since structural proteins tend to have relatively extended lifetimes (they usually serve a useful function in all cell types), they would be less likely to contain PEST sequences than regulatory proteins.

20. An amber mutation in a critical place in a protein sequence will not necessarily be suppressed by any suppressor tRNA, especially if only specific amino acids are tolerated at that position. The observation that an *su1* (serine-inserting) suppressor can suppress an amber mutation in the *trp* biosynthetic operon indicates that the amino acid serine is functional at the position in the protein which was mutated to an amber codon. The mutation in the arabinose operon, however, is at a position which will not accommodate serine and still give a functional protein. A glutamine-inserting suppressor will complement the arabinose mutation, but not the tryptophan operon.

16. (continued)

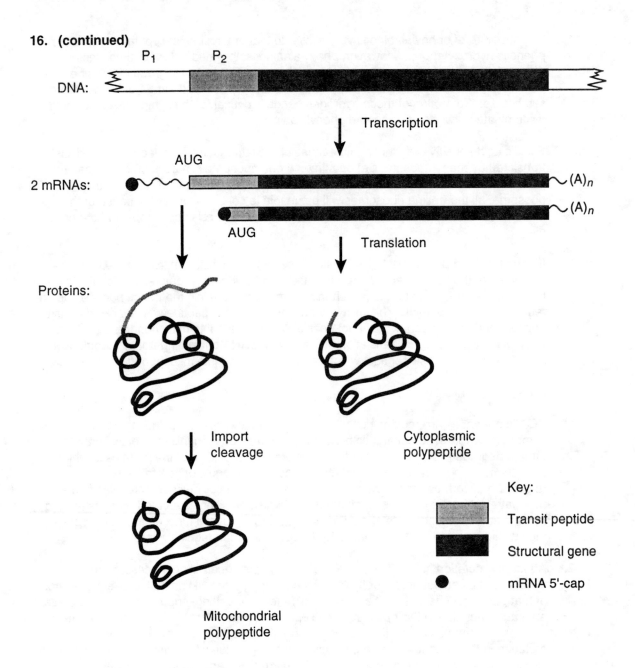

35 Regulation of Gene Expression in Prokaryotes

Chapter 35 Answers

2. **a.** Cells with the genotype $i^s o^+ z^+$ carry a super-repressor i^s mutation, which causes the repressor to be insensitive to inducer. The mutant repressor binds tightly to its operator under all conditions, and even though the operator and β-galactosidase loci are wild-type, no β-galactosidase would be produced in this strain. On media containing X-gal, with or without IPTG, colonies will be white.

   b. Cells with the genotype $i^s o^c z^+$ carry the same super-repressor as in (a), but the β-galactosidase gene is under the control of a constitutive operator, o^c. This mutation interferes with the ability of the repressor to bind and inhibit transcription. Therefore, β-galactosidase should be produced under all conditions (constitutively). On X-gal, with or without IPTG, colonies will be blue.

   c. Merodiploids with the genotype $i^s o^c z^- / i^+ o^+ z^+$ should behave like the strain described in part (a) above. The constitutive operator, o^c, is insensitive to the super-repressor, but it does not drive constitutive synthesis of β-galactosidase because it is in *cis* with an inactive β-galactosidase gene (z^-). The operator which controls a wild-type *lacZ* gene is wild-type (o^+), and is therefore sensitive to the super-repressor synthesized from the i^s allele. Recall that the repressor is a diffusible gene product, and can act in *trans*, unlike the operator.

   d. Merodiploids with the genotype $i^+ o^c z^- / i^- o^+ z^+$ would exhibit wild-type regulation of β-galactosidase activity, because there is one functional copy of wild-type repressor (i^+) and the only functional copy of the β-galactosidase gene (z^+) is under the control of a wild-type operator (o^+). β-galactosidase activity will be repressed in the absence of inducer when the repressor is bound to the operator. On X-gal alone, colonies will be white. In the presence of inducer, the β-galactosidase gene will be induced: on X-gal and IPTG, colonies will be blue.

4. a. A mutation in *lacO*$_1$ which alters one of the bases recognized by *lac* repressor would bind repressor more weakly than wild-type, and would therefore be more easily inducible. Depending on the residual affinity between operator and repressor, *lac* operon transcription would be inducible by lower than normal concentrations of IPTG, or might even be constitutive. The *lac* operon would still be subject to glucose repression.

    b. A mutation in *lacI* which abolishes its ability to bind allolactose, but does not affect DNA binding, would result in a super-repressor. It would bind its operator, inhibiting transcription, and could not be induced under any conditions. In these cells, *lac* operon transcription would only occur at basal levels.

    c. A mutation in the *lacI* Shine-Dalgarno sequence which improves its complementary to the 3'-end of 16S rRNA would result in increased levels of the *lacI* gene product, lac repressor. In order to saturate the increased levels of repressor with inducer, much higher levels of inducer would be required than for the wild-type *lacI* gene. Practically speaking, it is unlikely that intracellular levels of allolactose would ever be sufficiently high to observe expression of β-galactosidase in this strain when grown on lactose. *lac* operon transcription could be induced by growth on the more stable analog of allolactose, IPTG.

    d. A mutation which generates a nonsense codon in the *lacA* gene should have no effect on transcription of the *lac* operon.

6. It is generally accepted that the extent of homology of a promoter to the consensus −35 and −10 sequences is an important determinant of promoter strength. It might, therefore, be reasonably expected that genes for which transcripts are needed in great amounts would possess promoters with sequences very close to the consensus. Although this is in some cases true, it is not generally the case. The explanation for the less than perfect match of most promoters to the consensus sequence is to be found in the need to regulate transcription. Transcriptional regulation is achieved in many instances by the selective improvement of the affinity of specific promoters for RNA polymerase. Such selective improvement is well illustrated in the case of regulation of the *lac* operon. The *lac* promoter is by all accounts a rather weak one; its match to the consensus sequence is as follows:

	−35	−10
Promoter consensus sequence	TTGACA	TATAAT
Wild-type *lac* promoter	TTTACA	TATGTT
*lac*UV5	TTTACA	TATAAT

The strength of the wild-type *lac* promoter, and hence the level of expression of the *lac* operon, may, however, be significantly increased as a result of binding of the catabolite activator protein (CAP) upstream of the -35 region. This is because of the additional affinity of the promoter for RNA polymerase provided by the latter's interaction with CAP. A mutant containing two base replacements in the -10 region of the *lac* promoter (UV5), which dramatically improves the homology of the promoter to the consensus sequence, increases its inherent strength to such a degree that it no longer requires CAP binding for high-level transcription. As a consequence, expression of the *lac* operon in this mutant is no longer subject to regulation by catabolite repression, a potential drawback when both glucose and lactose are available as carbon sources.

8. The *ilv* operon encodes enzymes required for the biosynthesis of three amino acids, isoleucine, leucine and valine, and is therefore regulated in response to all three of these amino acids. An attenuation mechanism is used, based on the sequence of the leader peptide, which contains a central region in which 14 of 17 codons specify ile, leu or val. When levels of ile-tRNA, leu-tRNA or val-tRNA are low, the ribosome stalls in this central region of the leader peptide, allowing RNA polymerase to proceed through the downstream genes in the operon.

10. rRNA levels in wild-type *E. coli* (●), *rel*A mutant (∆), and *spoT* mutant cells (■) are shown below under normal conditions, after amino acid starvation and after readdition of amino acids.

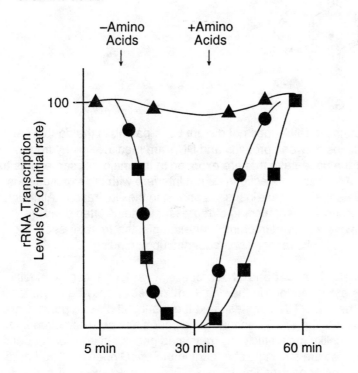

12.   Arg69 interacts with the face of guanine that is exposed in the major groove as shown below:

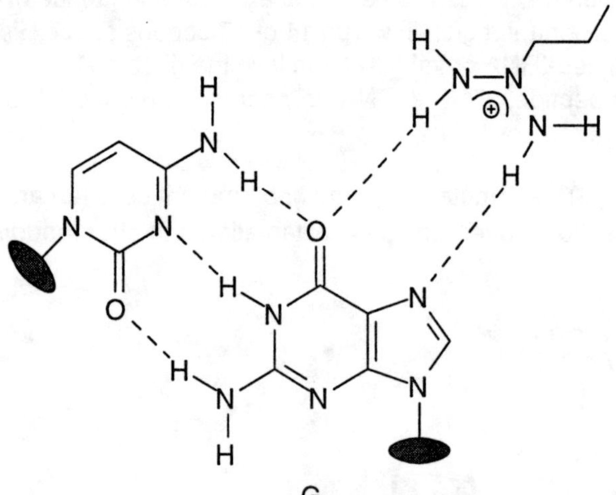

G_9

Binding of repressor proteins to DNA does not disrupt base-pairing in the double helix because specific interactions between proteins and DNA are mediated by hydrogen bonds with the edges of the base-pairs that are exposed in the major (or sometimes the minor) groove of the DNA.   These interactions do not interfere with the simultaneous formation of standard Watson-Crick base-pairs, as shown in the illustration. Although the DNA base-pairs are not disrupted, binding of repressor proteins often does introduce changes in the overall DNA structure. For example, the *trp* repressor introduces a noticeable bend in its recognition sequence upon binding.

14.   In addition to transcriptional differences and RNA processing, a poor ribosome binding site and a preponderance of rare codons in the *dnaG* coding region may be responsible for the discoordinate expression of DNA primase and the σ subunit of RNA polymerase. Indeed, the region upstream of the *dnaG* gene initiation codon has poor complementary to the 3'-end of 16S rRNA, unlike most efficiently translated genes. In addition, codons which are rare in *E. coli* genes are strikingly abundant in the *dnaG* open reading frame compared to the *rpoD* and *rpsU* open reading frames. For example, the AUA codon for isoleucine and the CGA, CGG, AGA and AGG codons for arginine, which are rarely used in *E. coli*, represent 32% and 23%, respectively, of the isoleucine and arginine codons in the *dnaG* gene. These codons are not represented among the 44 isoleucine codons and 46 arginine codons in the *rpoD* gene.

16.   A *cI⁻cro⁻* double mutant should be incapable of lysogeny because it cannot prevent the expression of the cascade of genes required for the lytic cycle. The absence of the *cro* gene function will not inhibit the lytic cycle as it usually does since the mutant is also *cI⁻*. This double mutant confirms that the lytic cycle is the "default" pathway for lambda phage.

36   Regulation of Gene Expression in Eukaryotes

Chapter 36 Answers

2.   No, this construct would not be activated by binding of *lexA* because *lexA* does not have a eukaryotic activation domain. This control experiment demonstrates that it is not simply binding of a protein to the upstream DNA which activates transcription of GAL1, further confirming the bifunctional nature of GAL4 protein and the necessity for protein-protein interactions. In other words, GAL4's DNA-binding domain is exchangeable with any DNA-binding domain whose recognition sequence can be introduced upstream of GAL1. However, GAL4p's transcription activation function requires a *bona fide* eukaryotic transcription activation domain.

4.   Promoter deletion studies are typically carried out using a recombinant vector with the promoter of interest attached to the DNA coding for an easily measured gene product (a so-called "reporter gene"). *GAL1* promoter-reporter gene fusions would be introduced into yeast cells and transcription of the reporter gene would be measured in the presence and absence of galactose. (In fact, since the *GAL1* UAS is bi-directional, a reporter gene can be attached to either the *GAL1* or the *GAL10* transcription initiation sites in order to precisely identify the galactose control region.) The figure below shows the predicted effect of deleting the *GAL1* upstream region to varying extents. The number of "+" symbols indicates relative transcript abundance. Note that a partial deletion of the GAL4 binding sites in the UAS (the second deletion in the series) is still galactose inducible, but not to the same extent as the construct with an intact UAS. This is because of cooperative binding of GAL4 to its binding sites: the more binding sites there are, the more tightly GAL4 is bound and the better it functions as a transcriptional activator.

4.   (continued)

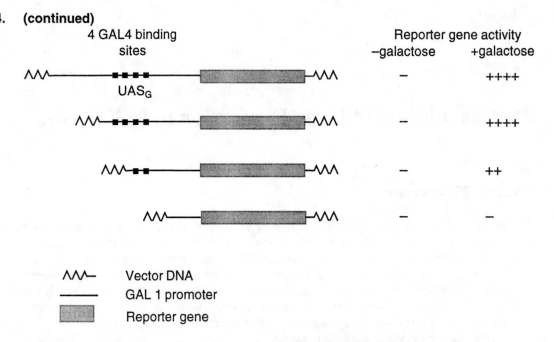

	Reporter gene activity	
	−galactose	+galactose
4 GAL4 binding sites / UAS$_G$	−	++++
	−	++++
	−	++
	−	−

∧∧∧−   Vector DNA

―――   GAL 1 promoter

▨   Reporter gene

6.   In wild-type haploid yeast, only one of the sets of mating type genes (α or a) are expressed at any one time. This exclusivity of expression is conditioned by a transposition mechanism that removes either set of genes from an inactive "storage" chromosomal context (*HMLα* and *HMRa* loci) to an active one (the *MAT* locus). The inactivity of *HMLα* and *HMRa* genes at their "storage" loci is due to the close proximity of "silencer" elements (*HMLE* for *HMLα* and *HMRE* for *HMRa*) that repress transcription of surrounding genes. A deletion of *HMLE* should remove the element repressing transcription of *HMLα* at the "storage" location, and result in the constitutive expression of *HMLα* genes from this locus.

The consequences of constitutive expression of *HMLα* are predicted to depend on the cell type. The diploid would sporulate normally. In haploid a cells, which have *HMLa* at the *MAT* locus (*MATa*), the cell should resemble the diploid, expressing both a and α genes simultaneously, and thus be unable to mate. (Reality check: in fact, *HMLE* deletion does not result in constitutive expression of the *HMLα* genes, because there is another silencer adjacent to the *HML* locus. Deletion of both silencers does give the phenotypes described in this paragraph. This disagreement between predicted and expected results demonstrates why it is important to perform experiments such as the deletion experiment described here.) With *HMLα* at the *MAT* locus (*MATα*), haploid α cells lacking *HMLE* would exhibit the behavior characteristic of the α mating type, except that the cells would become sterile at a rate conditioned by the frequency of transposition of *HMRa* to the *MAT* locus.

The α2 gene plays a dual role in mating type regulation: in haploid α cells, it represses a-specific genes, while in the diploid, it forms a heterodimer with a1 protein, which represses haploid-specific genes. Deletion of the α2 gene from *MATα* would result in expression of both α- and a-specific genes in the haploid cell. This cell would most likely be sterile. A *MATa/MATα2* diploid would be unable to sporulate because of its inability to repress haploid-specific genes. Note that upon transposition of *HMRa* to the *MAT* locus, the α2 deletion would be lost, and in subsequent rounds of mating type interconversions, the α2 allele present in the original strain at the *HMLα* locus would restore the α2 gene to *MATα*.

6. **(continued)**
   To summarize:

Deletion:	HMLE	MAT α2	Nominal mating type:
Phenotype:	α	sterile (restored to functional α upon mating type interconversion	MATα
	sterile (no restoration)	a	MATα
	normal	unable to sporulate	Diploid

8. The analog 5-azacytidine has a nitrogen at the 5-position of the otherwise normal pyrimidine ring. This analog therefore cannot be methylated at the 5-position, although its ability to form Watson-Crick base-pairs is unaffected. When cells are treated with 5-azacytidine, this nucleoside analog of cytidine is converted to the corresponding nucleotide via salvage pathways, and then incorporated into DNA. After a number of rounds of replication, the incorporation of 5-azacytidine results in hypomethylated regions in the DNA. Expression of some genes is inhibited by DNA methylation in their promoter regions; when these regions become hypomethylated, these genes can be transcribed and expressed.

10. Transcriptionally active chromatin is usually more swollen that inactive chromatin. In addition, the following characteristics have been used to differentiate between active and inactive chromatin in eukaryotic cells:
    a. Active chromatin is more susceptible to DNase I degradation than inactive chromatin. This can be shown for genes that are active only in certain tissues, and inactive in others. The regions that are more susceptible to enzyme digestion usually encompass a large domain around a transcriptionally active gene or cluster of genes. A particular gene expressed in a given tissue can be shown to be sensitive to DNase I digestion in that tissue, but not in a tissue where it is not expressed. For example, the globin genes are sensitive to DNase I digestion in chromatin isolated from erythrocytes, but not in chromatin isolated from oviduct.
    b. Active chromatin is often found to be associated with modified histones. Histones have been found to be subject to methylation and acetylation, and to phosphorylation in the case of histone H1. Histone H2A can also be found associated with ubiquitin. Ubiquitinated H2A is found preferentially in active chromatin. Such modifications have not been directly correlated with other characteristics of active chromatin, such as DNase I sensitivity.

10. (continued)

    c. High mobility group (HMG) proteins are found preferentially associated with active chromatin. It has been shown that HMG proteins can impart DNase I sensitivity to chromatin in reconstitution experiments. When chromatin from erythrocytes was stripped of its HMG proteins, globin genes lost their preferential sensitivity to DNase I digestion. Reconstitution with HMG proteins (which could be isolated from any tissue, including those not expressing globin genes) caused the globin genes to regain their susceptibility to DNase I digestion. When chromatin from brain (which does not express the globin genes) was reconstituted with HMG proteins, no sensitivity of the globin genes to DNase I digestion was observed. It thus appears as if HMG proteins will bind and confer sensitivity to DNase I digestion to chromatin packaged in a state different from that of inactive chromatin, perhaps due to the binding of specific regulatory factors. HMG proteins do bind to transcriptionally <u>in</u>active chromatin in such reconstitution experiments, but do not confer DNase I sensitivity upon it.

    d. DNA methylation correlates with inactive chromatin. Most cytosine residues in CG dinucleotides are found to be methylated in vertebrate DNA. In some cases, it has been shown that tissue-specific or developmentally regulated genes have methylated CG residues at times during development when, or in tissues where, they are not expressed, and become demethylated (or hypomethylated) when they are expressed.

12. The portion of the tropomyosin protein that interacts with actin must be conserved in all of the tropomyosin isoforms, and therefore is presumably found in one or more of the exons that are constant in different tissues. The exons encoding amino acids 81-125, 126-164, 165-188, 214-234, and 235-257 are candidates for actin-binding sequences. Exons which are not included in all tropomyosin isoforms might encode parts of the protein with tissue-specific functions. For example, in striated muscle, an interaction between tropomyosin and troponin T is mediated by amino acids encoded in the exons denoted 189-213 and 258-284 in figure 36.24.

14. When cells resistant to high levels of methotrexate are assayed for dihydrofolate reductase (DHFR), the levels of this enzyme are elevated about 1,000-fold. Large amounts of dihydrofolate reductase are made in these resistant cell lines by amplifying the DHFR gene, generating a large gene dose that produces a correspondingly large amount of DHFR mRNA. The large amount of mRNA is translated to produce high levels of DHFR enzyme.

Methotrexate binds very tightly to dihydrofolate reductase and normally would be very toxic to rapidly growing cells. When large amounts of enzyme are made, all of the cellular methotrexate is bound, and the remaining enzyme activity is sufficient to regenerate tetrahydrofolate from dihydrofolate and allow cell growth.
The copy number of the dihydrofolate reductase gene could be determined by probing with a specific DNA sequence made using recombinant techniques or synthetic DNA made on an oligonucleotide synthesizer. (The "dot blot" procedure would be one of the easiest methods of hybridization analysis.) DNA isolated from normal and methotrexate-resistant cells would be hybridized to the labeled probe, and the amount of hybridizing DNA would be compared to that hybridizing to a probe corresponding to a gene which is not expected to be amplified. The DNA from the resistant cells would bind more of the DHFR probe relative to the control than the DNA isolated from normal cells.

16. One strategy that is employed to generate sufficient amounts of a product required at a specific stage of development is gene amplification. An example of this strategy is to be found in the sea urchin, which achieves a high rate of histone synthesis during embryogenesis by producing a large number of copies of the histone genes. Another example of gene amplification is provided by *Xenopus laevis*. The amounts of ribosomal RNA in frog eggs can be greatly increased by amplification of the rRNA genes. The amplified DNA is extrachromosomal, located near the nucleolus, transcribed during oogenesis, and subsequently degraded. The 5S rRNA genes of *Xenopus laevis* fall into two families. There are approximately 20,000 copies of a 5S rRNA gene expressed during oogenesis, and about 400 copies of a 5S rRNA gene expressed in somatic cells and growing oocytes. The expression of 20,000 copies of the oocyte gene provides for very high levels of 5S rRNA during oocyte growth.

18. The radioactively labeled cloned DNA fragment would be hybridized to giant chromosomes much like those illustrated in Figure 36-xx. The probe could be radioactively labeled, or nonradioactive nucleic acid labeling methods can be used. After washing off the unhybridized probe, autoradiography of the hybridized chromosomes would reveal which band hybridized to the cloned DNA.

The availability of these polytene chromosomes greatly facilitates the process of gene mapping in *Drosophila*. Each giant chromosome in the insect salivary gland actually represents many copies of the genomic DNA, and will therefore hybridize many copies of the labeled probe. This amplification in signal greatly facilitates detection of the hybridizing band. Similar experiments can be performed using human chromosomes. However, because each human chromosome has only one copy of the genomic DNA, highly sensitive and technically difficult detection methods must be employed to detect specific hybridization to human chromosomes.

37   Immunobiology

Chapter 37 Answers

2.   It has been known for a long time that if the thymus is removed from animals, they show a reduced reaction to foreign antigens. Thus, the T cells must be important in mediating the B cell response. Helper T cells stimulated by a specific antigen present that antigen to B cells in a concentrated form. In B cells with the appropriate antigen specificity, this focusing of the antigen causes the immunoglobulin (Ig) membrane receptors to cluster together on the B cell membrane, forming what is called a CAP structure. CAP formation triggers proliferation of the B cells. When the thymus is removed, these animals are depleted in T cells and the focusing of the antigen does not occur (very few helper T cells are left).

4.   An epitope is the antigenic structure in the presence of which a given antibody-producing cell responds and produces antibodies of a specific structure (the idiotype). In serum of polyclonal origin, there should be antibodies directed against many of the antigen's epitopes.   Anti-idiotypic antibodies are antibodies which react with the variable region of these antibodies. Since the variable region contains the antigen-binding site, some of the anti-idiotypic antibodies will recognize those sequences of the eliciting antibody that would be in contact with antigen. In a phenomenon known as "molecular mimicry", it has been shown that some anti-idiotypic antibodies mimic the original antibody-antigen interaction. Thus, the anti-idiotypic antibody carries in its variable region a sequence or structure which closely resembles the epitope with which the original antibody interacted. In schematic form:

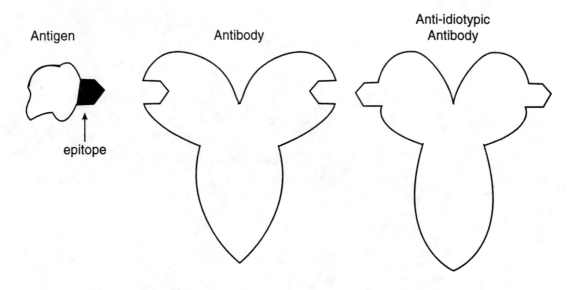

Antigen   Antibody   Anti-idiotypic Antibody

epitope

6. The efficiency of the polymerase chain reaction (PCR) in amplifying DNA fragments generally decreases as the distance between primers increases. It is likely that in unrearranged embryonic DNA, the L and C portions of the light chain genes are too far apart to generate a PCR product under normal conditions. In the rearranged genomic DNA, the proximity of L and C regions will permit amplification by PCR of a fragment containing 5'-L, rearranged V-J, and 3'-C segments, separated by two introns.

8. The mouse from strain A could be irradiated with whole-body irradiation to destroy the immune system. Then the bone marrow and thymus from strain B could be transplanted into the irradiated mouse, giving it a competent immune system that is isogeneic with strain B tissue. If you were working with newborn mice, you could inject cells from strain B mice into strain A mice. This would make the strain A mice tolerant to transplants from strain B when they reach adulthood.

10. Antibodies with catalytic activity have been produced by designing an analog of the transition-state intermediate as antigen. Some enzymes work by binding substrates and stabilizing the transition-state intermediate, thus facilitating the conversion of substrate to product. If a transition state analogue is bound to a protein like albumin, and used to generate antibodies, some of the resulting antibodies will bind to the desired substrate in such a way as to favor formation of the transition state. Such antibodies will have catalytic activity, in some cases on a par with biological enzymes. With these techniques, enzymes can be designed to catalyze reactions that have no known natural counterparts.

12. A high concentration of a thiol compound such as 2-mercaptoethanol would reduce the disulfide bonds holding the heavy chains together and holding the light chains to the heavy chains, allowing the dissociation of the chains. The reduced antibody would not have the necessary structure to bind antigen.

38   Cancer and Carcinogenesis

Chapter 38 Answers

2. Ethionine is taken up by cells and metabolized to the S-adenosylmethionine (SAM) analog, S-adenosylethionine (SAE). SAE is a substrate analog for various methyltransferases and inhibits cellular methylation. It is thought that this inhibition causes a change in the methylation pattern of DNA, resulting in hypomethylation of the 5'-flanking sequences of some genes. Many genes involved in development are expressed when their 5'-flanking regions are hypomethylated. When expressed at the wrong time during development, these genes may reverse the control of cellular growth which is the hallmark of normal differentiated cells. While ethionine is a nonmutagenic carcinogen (it does not lead to heritable changes in the DNA sequence), it does change methylation patterns, which can affect gene expression and lead to cancer. Interestingly, the carcinogenic effects of ethionine can be prevented by administering methionine.

4. If the oncogene *myc* functions ubiquitously in cell proliferation, this may explain why it is found associated with a number of different types of tumors. In other words, overexpression of the *myc* oncogene in nearly any cell would have the same result because all cells respond to this protein in the same fashion. The oncogene *sis* acts as a growth factor related to PDGF. Only cells expressing the PDGF cell surface receptor will respond to the *sis* gene product, explaining why this oncogene is found only in cancers simultaneously expressing the PDGF receptor.

6. Cellular protooncogenes can be converted to oncogenes by a number of mechanisms. The gene product can be made in higher than normal amounts (*src, jun, myc*), or normal amounts can be made but the protein altered so that it behaves differently (*ras*). The *sis* oncogene is similar in amount and structure to the cellular gene but its product is expressed in the wrong cells. Any one of these alterations can result from point mutations, DNA rearrangements including deletions, duplications, and chromosomal translocations, or retroviral activation.

8. The HIV virus can destroy the immune system by killing the T4 cells, which are central to the immune response. Some cancer cells can be detected by the immune system as foreign and targeted by killer T cells. If the patient's immune system cannot detect these cancer cells, the cells will grow and spread. Some of the rare cancers that develop in AIDS patients may therefore develop because these patients lack immune surveillance.

39   The Human Immunodeficiency Virus (HIV) and Acquired Immunodeficiency Syndrome (AIDS)

Chapter 39 Answers

2. Life cycle differences between retroviruses are in part due to their differing host cells, and in part due to the complexity (or lack thereof) of the different genomes. Most retroviruses require only three genes for their replication, and it is surprising that HIV-1 has six additional open reading frames (table 39.2). Two of these, *tat* and *rev*, are essential, while the remaining four do not seem to be absolutely necessary *in vitro*. Because the sequences of all six of these extra genes are highly conserved among the immunodeficiency viruses, it is believed that their protein products do play important roles in the infectivity and the life cycle of this class of retroviruses. For example, *rev* protein functions in a post-transcriptional pathway to promote the expression of late genes. Other retroviruses lacking the *rev* protein must rely on different mechanisms to regulate the latency of infection. These accessory proteins may represent targets for therapeutic intervention in AIDS and HIV-positive patients.

4. The pathway for the conversion of AZT to its triphosphate derivative involves the salvage enzyme thymidine kinase, followed by action of the enzymes deoxythymidylate kinase and NDP kinase. The resulting azidothymidine triphosphate is incorporated into the growing DNA chain, bringing replication to a halt, as shown below.

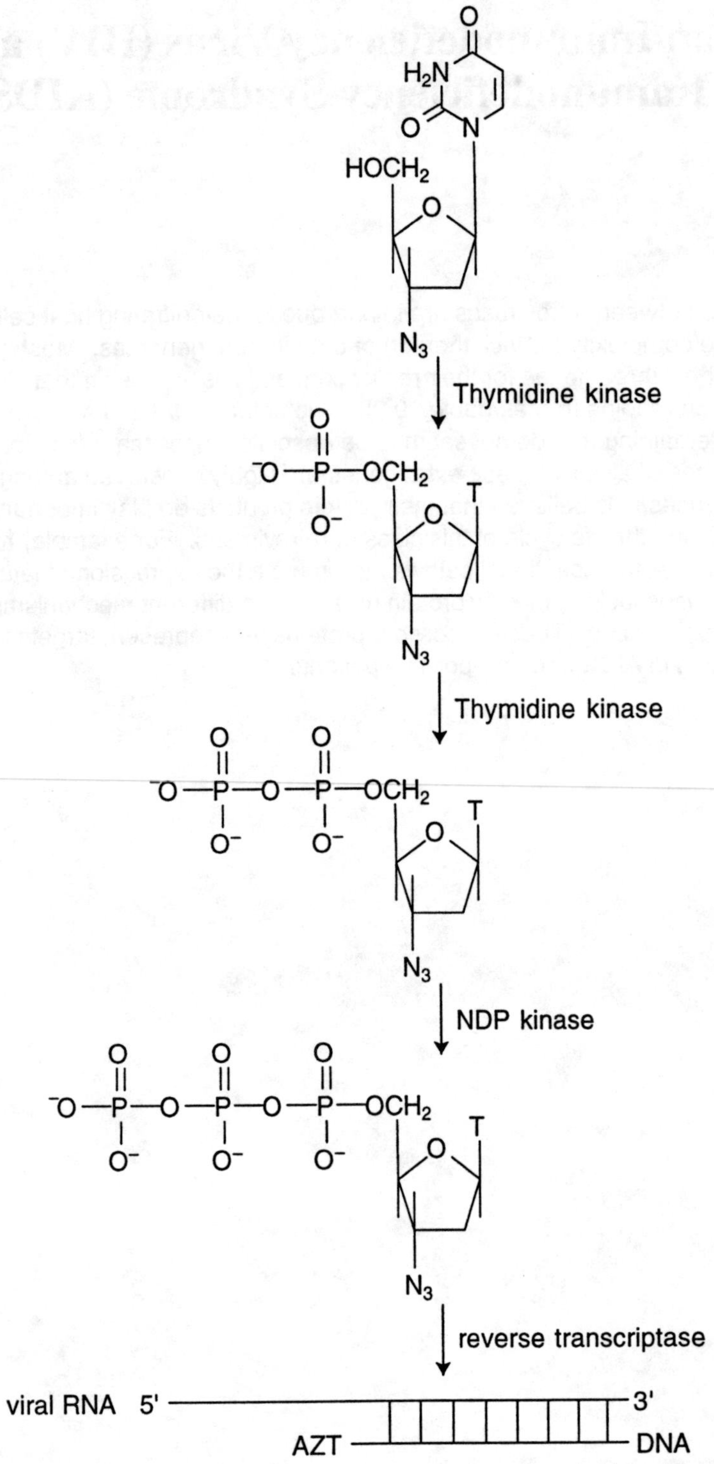

6.   Retroviruses generally exhibit a higher mutation frequency that the host genomes because their life cycle includes an RNA intermediate. This RNA intermediate is generated by RNA polymerase II, which is much less accurate than DNA polymerase, and is thus likely to introduce occasional mutations into the viral RNA. The RNA intermediate is converted to double-stranded DNA through the action of reverse transcriptase, which also lacks proof-reading activity, and is therefore also likely to introduce mistakes into the viral genome.

8.   The inhibition of translation by nontranscribed sequences in a eukaryotic system is surprising because of the spatial separation between transcription and translation. That is, once the mRNA is transcribed and processed, it should be exported and translated efficiently regardless of the DNA sequences flanking the transcribed region. In this case, it may be that the upstream region helps recruit factors which bind the mRNA and facilitate its efficient export or translatability. This is yet another example of the surprising complexity of HIV-1 gene regulation.

TEST ITEM FILE

Prepared by

Hugh Akers
Cindy Klevickis

Boston   Burr Ridge, IL   Dubuque, IA   Madison, WI   New York   San Francisco   St. Louis
Bangkok   Bogotá   Caracas   Lisbon   London   Madrid
Mexico City   Milan   New Delhi   Seoul   Singapore   Sydney   Taipei   Toronto

CONTENTS

Chapter 1 Cells, Organelles and Biomolecules

1. Of the following pairs of cell types, the most substantial difference is between
   a. animal cells and plant cells.
   b. prokaryotic cells and eukaryotic cells.
   c. cells that have cell walls and those that do not have cell walls.
   d. yeast cells and tetrahymena protozoa.
   answer: b

2. All cells have
   a. a nucleus and a cell membrane.
   b. a cell wall and a cell membrane.
   c. a nucleus and mitochondria.
   d. a cell membrane.
   answer: d

3. In plant and animal cells, the three-dimensional integrity of the cell is maintained, in part, by an internal network of filaments called
   a. the cell wall.
   b. the Golgi apparatus.
   c. the cytoskeleton.
   d. the endoplasmic reticulum.
   answer: c

4. Of the following, which are prokaryotes?
   a. viruses
   b. bacteria
   c. yeast
   d. plant cells
   answer: b

5. Which of the following classes of molecules comprises the largest percentage of the dry weight of a typical bacterial cell?
   a. lipids
   b. inorganic ions
   c. amino acids, sugars and other small molecules
   d. macromolecules (proteins, nucleic acids and polysaccharides)
   answer: d

6. Most cells contain about _____% water by weight.
   a. 30
   b. 50
   c. 70
   d. 90
   answer: c

7. Which of the following classes of molecules tend to be poorly soluble in water?
   a. sugars
   b. amino acids
   c. inorganic ions
   d. fatty acids
   answer: d

8. Large carbohydrate molecules are called
   a. polysaccharides.
   b. polypeptides.
   c. polybutyrates.
   d. polynucleotides.
   answer: a

9. Phospholipids spontaneously assemble to form micelles or bilayers in water solution through the formation of
   a. hydrophobic and hydrophilic interactions.
   b. covalent bonds.
   c. interactions with membrane proteins.
   d. ionic bonds.
   answer: a

10. When amino acids are linked together to form a linear chain, the byproduct is
    a. a salt.
    b. a carboxylic acid.
    c. a base.
    d. water.
    answer: d

11. In the globular, three-dimensional structure of a typical enzyme in water solution, polar, hydrophilic amino acid side chains are predominantly found
    a. in the interior of the molecule.
    b. on the exterior surface of the molecule.
    c. randomly distributed throughout the molecule.
    d. in contact with the cell membrane.
    answer: b

12. Enzymes regulate the reactions that occur within living cells by
    a. providing the energy for these reactions to occur.
    b. accelerating the rate of the particular cell reactions that they catalyze.
    c. making cell reactions more thermodynamically favorable.
    d. providing the raw materials for cell reactions.
    answer: b

2

13. The biochemical reactions that result in the breakdown of organic compounds to simpler substances are called
    a. synthetic reactions.
    b. anabolic reactions.
    c. catabolic reactions.
    d. ionic reactions.
    answer: c

14. This functional group is called

    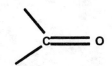

    a. an hydroxyl group.
    b. an amino group.
    c. a carboxylic acid group.
    d. a carbonyl group.
    answer: d

15. The products of a chemical reaction between an alcohol and a carboxylic acid are
    a. an ether and water.
    b. a salt and water.
    c. a diol and acid.
    d. an ester and water.
    answer: d

16. When an amine reacts with a carboxylic acid, the major product is
    a. an amide.
    b. a base.
    c. a salt.
    d. ammonia.
    answer: a

17. The two chains of a DNA double helix are held together by
    a. covalent bonds.
    b. ionic bonds.
    c. hydrogen bonds.
    d. hydrophilic interactions.
    answer: c

18. A class of small molecules that typically forms zwitterions in water solution is
    a. sugars.
    b. fatty acids.
    c. amino acids.
    d. nucleic acids.
    answer: c

19. According to fossil evidence, life on earth arose
    a. about 3-4 billion years ago.
    b. about 1 billion years ago.
    c. about 1 million years ago.
    d. about 100,000 years ago.
    answer: a

20. By comparing nucleic acid sequences between representative organisms for all of the different kinds of life on earth, Carl Woese developed an evolutionary tree that has these three main branches:
    a. protists, fungi, bacteria.
    b. animals, plants, fungi.
    c. bacteria, fungi, higher eukaryotes.
    d. archaebacteria, eubacteria, eukaryotes.
    answer: d

21. Energy for many biochemical reactions comes from
    a. enzymes.
    b. DNA.
    c. ATP.
    d. phospholipids.
    answer: c

22. Each of these subcellular structures is enclosed by a membrane EXCEPT
    a. the nucleus.
    b. ribosomes.
    c. mitochondria.
    d. chloroplasts.
    answer: b

23. The four most abundant elements in living cells are
    a. C, H, O, N
    b. C, O, P, S
    c. C, O, H, Ca
    d. C, O, N, P
    answer: a

24. This compound is

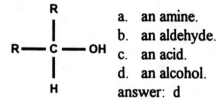

    a. an amine.
    b. an aldehyde.
    c. an acid.
    d. an alcohol.
    answer: d

25. In water solution, amines usually
    a. loose a proton to become negatively charged.
    b. add a proton to become positively charged.
    c. undergo dehydrogenation to become imines.
    d. hydrolyze to form ketones.
    answer: b

26. Plant cell walls are made of
    a. phospholipids
    b. linear polysaccharides
    c. proteins
    d. lipids and proteins
    answer: b

27.

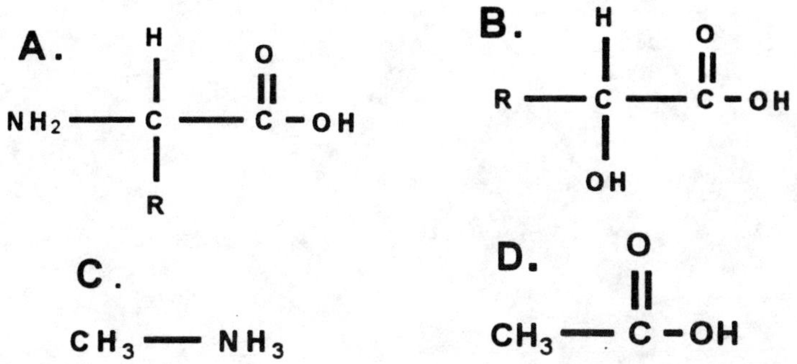

A.

$$NH_2 - \underset{\underset{R}{|}}{\overset{\overset{H}{|}}{C}} - \overset{\overset{O}{||}}{C} - OH$$

B.

$$R - \underset{\underset{OH}{|}}{\overset{\overset{H}{|}}{C}} - \overset{\overset{O}{||}}{C} - OH$$

C.

$$CH_3 - NH_3$$

D.

$$CH_3 - \overset{\overset{O}{||}}{C} - OH$$

Of the four structures shown above, which represents an amino acid?
    a. A.
    b. B.
    c. C.
    d. D.
    answer: a

28. The number of different cell types in a human is
    a. about 20.
    b. about 100.
    c. about 100,000.
    d. about 10^{14}.
    answer: b

29. Each of the following subcellular structures contains DNA EXCEPT
    a. the nucleus.
    b. ribosomes.
    c. chloroplasts.
    d. mitochondria.
    answer: b

30. In the double-helical structure of DNA, the phosphate groups are
    a. negatively charged and on the outside of the molecule.
    b. positively charged and on the outside of the molecule.
    c. negatively charged and buried within the interior parts of the molecule.
    d. positively charged and buried within the interior parts of the molecule.
    answer: a

Chapter 2 Thermodynamics in Biochemistry

1. Which of the following can be classified as an <u>extensive</u> property of a substance?
    a. density
    b. pressure
    c. volume
    d. temperature
    answer: c

2. The First Law of Thermodynamics says that
    a. a chemical reaction will occur spontaneously only if it results in a net release of free energy.
    b. the total amount of energy in the universe is constant.
    c. the total amount of entropy in the universe is always increasing.
    d. a chemical reaction will proceed spontaneously only if entropy increases.
    answer: b

3. In most biochemical reactions, ΔPV is
    a. negligible
    b. positive
    c. negative
    d. dependent on ΔE
    answer : a

4. According to the Second Law of Thermodymnamics,
    a. the total entropy change in any reaction that occurs spontaneously must be greater than 0.
    b. the total entropy change in any reaction that occurs spontaneously must be less than 0.
    c. the total entropy of the universe remains constant.
    d. the total entropy of the universe is always decreasing.
    answer: a

5. Which of the following processes might result in an overall decrease in entropy?
    a. vaporization of a liquid
    b. melting of a solid
    c. mixing of two miscible solids in an ideal solution
    d. dissolution of salt in water
    answer: d

6. At equilibrium the free energy, ΔG
    a. depends only on the temperature.
    b. is positive.
    c. is 0.
    d. is negative.
    answer: c

7. Given R = 1.98 cal/degree • mole, and Keq = 10^2, calculate $\Delta G°$ at 25° (T = 298 K).
   a. 2.73 kcal/mole
   b. 2 kcal / mole
   c. 0 kcal/ mole
   d. -2.73 kcal/ mole
   answer: d

8. The reaction between glucose and inorganic phosphate to form glucose-6-phosphate has a $\Delta G°$ of +3.3 kcal/mole at 298 K. Which of the following statements is TRUE?
   a. This reaction will occur spontaneously only if the glucose-6-phosphate concentration is much greater than the glucose concentration.
   b. This reaction will occur spontaneously only if the inorganic phosphate concentration is very low.
   c. This reaction is at equilibrium.
   d. This reaction will not occur spontaneously under standard conditions.
   answer: d

9. $\Delta G°'$ is -0.05 kcal/mole for the reaction between acetyl-CoA and oxaloacetate to form citryl-CoA. What is Keq' at 298 K?
   a. about 0.05
   b. about 0.1
   c. about 1.0
   d. about 10
   answer: c

10. At pH 7 under physiological conditions, $\Delta G°$ is approximately -7.5 kcal/mole for the hydrolysis of ATP. This strongly favorable free energy results from
    a. a very favorable entropy change that overcomes an unfavorable enthalpy change.
    b. a very favorable enthalpy change that overcomes an unfavorable entropy change.
    c. a combination of favorable entropy and enthalpy changes.
    d. a favorable entropy change that overcomes the overall unfavorable effect on the   free energy of pH 7 compared to pH 0.
    answer: c

11. Most of the change in energy in chemical reactions is due to a change in
    a. vibrational energy.
    b. translational energy.
    c. electronic energy.
    d. rotational energy.
    answer: c

8

12. Calculate $\Delta G°$ for Glucose + ATP → Glucose 6P + ADP. **Note that:**

   Glucose + Pi → Glucose 6P $(\Delta G° = +3.3$ kcal/mole$)$

   ATP → ADP + Pi $(\Delta G° = -7.5$ kcal/mole$)$

   a. -10.8 kcal/mole
   b. -4.2 kcal/mole
   c. +4.2 kcal/mole
   d. +10.8 kcal/mole
   answer: b

13. ΔG is a measure of
   a. the total energy change for a reaction.
   b. the enthalpy change for a reaction.
   c. the total entropy change for a reaction.
   d. the free energy change for a reaction.
   answer: d

14. For a reversible isothermic process
   a. ΔS is independent of the absolute temperature.
   b. ΔS is directly proportional to the absolute temperature.
   c. ΔS is inversely proportional to the absolute temperature.
   d. ΔS is proportional to the square of the absolute temperature.
   answer: c

15. In the presence of a free radical initiator, ethylene polymerizes to form polyethylene. **Over the course of this reaction, the translational entropy**
   a. increases.
   b. remains the same.
   c. decreases.
   d. There is not enough information to determine the effect on the translational entropy.
   answer: c

16. Substrate binding to an enzyme frequently results in a positive ΔS. The reason for this is that
   a. there is an increase in the translational entropy of the substrate.
   b. there is a decrease in the entropy of the surrounding water molecules.
   c. water molecules are displaced when the substrate binds, resulting in an increase in entropy.
   d. the rotational entropy of the enzyme is increased.
   answer: c

17. Gibbs proposed that a reaction can occur spontaneously if and only if
   a. ΔG is 0.
   b. ΔG is negative.
   c. $T\Delta S$ is 0.
   d. $T\Delta S$ is negative.
   answer: b

9

18. In aqueous solution, Mg^{2+} is coordinated by 6 water molecules. Two to three of these molecules can be displaced by the phosphate groups of ATP to form Mg^{2+} ATP. If two water molecules are displaced, the product is bidentate Mg^{2+} ATP, whereas tridentate Mg^{2+} ATP is formed when three molecules are displaced. From the principles of thermodynamics, you can conclude that:
   a. in tridentate Mg^{2+} ATP, the Mg^{2+} binds more tightly to the ATP than it does in bidentate Mg^{2+}ATP because ΔS is greater.
   b. in bidentate Mg^{2+} ATP, the Mg^{2+} binds more tightly to the ATP than it does in tridentate Mg^{2+}ATP because ΔS is greater.
   c. the Mg^{2+} will bind equally tightly to the ATP in bidentate or tridentate Mg^{2+}ATP.
   d. water molecules bind more tightly to Mg^{2+} than do the ATP phosphates.
   answer: a

19. For a hypothetical reaction in which $HA \rightleftharpoons H^+ + A^-$, Keq is 10^{-1} at pH 0. What is Keq' (at pH 7)?
   a. 10^{-8}.
   b. 10^{-1}.
   c. 10^6.
   d. 10^8.
   answer: c

20. The presence of a catalyst,
   a. decreases ΔG.
   b. does not affect ΔG.
   c. increases ΔG.
   d. increases $\Delta G^{O'}$.
   answer: b

21. At pH 7, the overall charge of ATP is closest to
   a. 0
   b. -2
   c. -3
   d. -4
   answer: d

22. As the temperature increases,
   a. $T\Delta S$ becomes more positive, but ΔG remains the same.
   b. $T\Delta S$ becomes more negative but ΔG remains the same.
   c. ΔG becomes more positive.
   d. ΔG becomes more negative.
   answer: d

10

23. Which of the following is NOT a state function?
    a. work
    b. ΔS
    c. ΔG
    d. ΔH
    answer: a

24. The maximum amount of useful work that can be obtained from a reaction is equal to
    a. $T\Delta S$
    b. $-T\Delta S$
    c. ΔG
    d. $-\Delta G$
    answer: d

25. Consider the following reaction: Phosphocreatine $+ H_2O \rightarrow$ creatine $+ P_i$.
    At 37 ° C (310 K), $\Delta G°$ for this reaction is -10.2 kcal/mole. Assume that the actual
    concentrations of phosphocreatine, creatine and P_i are all 0.001 M. What is ΔG for this
    reaction?
    a. -4.2 kcal/mole
    b. -6.0 kcal/mole
    c. -10.2 kcal/mole
    d. -14.4 kcal/mole
    answer: d

26. During glycolysis, ATP synthesis is coupled to the conversion of phosphoenolpyruvate (PEP)
    to pyruvate according to the following chemical equation:

    $$PEP + H^+ + ADP \rightarrow Pyruvate + ATP \quad (G°' = -7.5 \text{ kcal/mole})$$

    Note that: $ATP + H_2O \rightarrow ADP + P_i \quad (G°' = -7.5 \text{ kcal/mole})$ What is the $\Delta G°'$ for the
    hydrolysis of PEP to form pyruvate $+ P_i$?
    a. +7.5 kcal/mole
    b. 0 kcal/mole
    c. -7.5 kcal/mole
    d. -15 kcal/mole
    answer: d

27. $G°'$ for a hypothetical reaction: $A \leftrightarrow B+C$ is 2.7 kcal/mole at 25° C. If the concentrations of
    both G and C are 0.01 M, calculate the minimum concentration of A for which this reaction
    would occur spontaneously.
    a. 10^{-4} M
    b. 10^{-2} M
    c. 0.1 M
    d. 1 M
    answer: b

11

28. For the reaction: A ⇌ B, there is 9% product and 91% reactant at equilibrium. What is $G°$?
    a. -1.36
    b. 0.1
    c. 1.36
    d. 10
    answer: a

29. $G°$ for the conversion of glycerol-3-phosphate to glycerol + P_i is -2.2 kcal/mole. Thus, you can conclude that
    a. this reaction will occur spontaneously.
    b. this reaction may or may not occur spontaneously, depending on the actual concentrations of reactants and products.
    c. this reaction will only occur spontaneously if it is coupled to another reaction that has a more favorable $G°$.
    d. this reaction is so strongly favorable that it could be coupled to drive an otherwise unfavorable cell reaction.
    answer: b

30. Consider the following reaction: A + H^+ ⇌ AH. If all other factors remain constant,
    a. the reaction will be more favorable at pH 7 than at pH 0.
    b. Keq will be less at pH 7, however, ΔG will remain the same.
    c. the reaction will be less favorable at pH 7 that at pH 0.
    d. the thermodynamics of the reaction will not be affected by pH, however the reaction will proceed more rapidly at pH 0 than at pH 7.
    answer: c

Chapter 3 The Structure and Function of Water

1. What is the pH of an aqueous solution that has a $[H^+]$ of 6.65×10^{-10} M?
    a. 9.18
    b. 4.82
    c. 5.47
    d. 1.89
    e. 0.82
    answer: a

2. What is the pH of an aqueous solution that has a $[OH^-]$ of 3.45×10^{-5} M?
    a. 3.47
    b. 4.47
    c. 5.53
    d. 9.54
    e. none of the above
    answer: d

3. A 400 mL solution contains 845 mg of sodium hydroxide. What is the pH of this aqueous solution?
    a. 3.28
    b. 1.28
    c. 12.7
    d. 10.7
    e. none of the above
    answer: c

4. A saturated solution contains 0.009 g of magnesium hydroxide per liter. What is the pH of this aqueous solution, assuming that the magnesium hydroxide is completely ionized.?
    a. 10.49
    b. 3.51
    c. 3.66
    d. 10.19
    e. 10.34
    answer: a

5. Blood plasma has a pH of 7.4. What is the $[H^+] / [OH^-]$?
    a. 4.0×10^{-8}
    b. 2.51
    c. 0.16
    d. 6.41
    e. 0.39
    answer: c

6. If in an aqueous solution the $3 [H^+] = [OH^-]$, what is the pH of the solution?
   a. 6.76
   b. 4.67
   c. 6.13
   d. 7.87
   e. 7.24
   answer: e

7. If in an aqueous solution the $[H^+] = 7.5 [OH^-]$, what is the pH of the solution?
   a. 6.56
   b. 6.97
   c. 7.44
   d. 7.03
   e. none of the above
   answer: a

8. A 0.01 M solution of a monoprotic organic acid has a pH of 3.67.   What is the pKa of the acid?
   a. 5.67
   b. 1.67
   c. 5.23
   d. 6.23
   e. none of the above
   answer: c

9. The pKa of chloroacetic acid is 2.85. What is the pH of a 0.05 M solution of this acid?
   a. 4.85
   b. 3.85
   c. 1.74
   d. 2.04
   e. none of the above
   answer: d

10. A monocarboxylic acid and its potassium salt were combined in a 2:1 mole ratio and dissolved in water. How might you mathematically define the pH of the resulting solution?
    a. pKa - 0.30
    b. pKa + 0.30
    c. pKa - 0.70
    d. pKa + 0.70
    e. 0.30
    answer: a

11. A 1.00 g sample of propionic acid and 0.86 g of potassium propionate were combined with enough water to make 0.10 L of solution. If the pKa of propionic acid is 4.87, what is the pH of the solution?
     a. 4.30
     b. 4.62
     c. 4.87
     d. 5.44
     e. none of the above
     answer: b

12. Sodium malate (2.22 g) and sodium hydrogen malate (0.73 g) were dissolved in water and diluted to 0.5 L   What is the pH of the buffer? Use pKa values of 3.4 and 5.1 for malic acid (2-hydroxybutanedioic acid)
     a. 5.5
     b. 4.7
     c. 3.8
     d. 3.0
     e. none of the above
     answer: a

13. At what pH's will the average charge on the phosphate species be: - 0.5, -1.0, -1.5? Use pKa values of 2.12, 7.21, and 12.7 for phosphoric acid.
     a. 2.12, 7.21, 12.7
     b. 2.12, 4.67, 7.21
     c. 4.67, 9.96, 12.7
     d. 7.21, 9.96, 12.7
     e. none of the above
     answer: b

14. What is the pH of a 500 mL solution that contains 200 millimoles of phosphoric acid and 130 millimoles of potassium hydroxide? Use pKa values of 2.12, 7.21, and 12.7 for phosphoric acid.
     a. 6.8
     b. 7.6
     c. 12.3
     d. 13.1
     e. none of the above
     answer: d

15. Phosphoric acid (1.47 g.) and 2.22 g of potassium phosphate were dissolved in water and diluted to 0.50 L. What is the pH of the solution? Use pKa values of 2.12, 7.21, and 12.7 for phosphoric acid.
     a. 1.63
     b. 6.50
     c. 6.70
     d. 7.70
     answer: c

15

16. Phosphoric acid (5.12 g.) and 4.13 g of potassium hydroxide were dissolved in water and diluted to 0.65 L. What is the pH of the solution? Use pKa values of 2.12, 7.21, and 12.7 for phosphoric acid.
    a. 7.37
    b. 7.07
    c. 1.90
    d. 7.30
    e. none of the above
    answer: b

17. What is the pH of a 250 mL solution that contains 0.75 g of potassium hydroxide and 4.76 g of potassium dihydrogen phosphate? Use pKa values of 2.12, 7.21, and 12.7 for phosphoric acid.
    a. 2.54
    b. 1.70
    c. 7.42
    d. 7.00
    e. none of the above
    answer: d

18. Which of the following is true?
    a. Intracellular and blood plasma pH's are buffered with both bicarbonate and phosphate systems
    b. Intracellular pH is buffered with bicarbonate systems and blood plasma pH is buffered with phosphate systems
    c. Intracellular pH is buffered with phosphate systems and blood plasma pH is buffered with bicarbonate systems
    d. Intracellular and blood plasma pH's are not buffered with bicarbonate or phosphate systems
    answer: c

19. What is the pH of a 250 mL solution that contains 3.90 g of potassium monohydrogen phosphate and 3.17 g of potassium dihydrogen phosphate? Use pKa values of 2.12, 7.21, and 12.7 for phosphoric acid.
    a. 7.41
    b. 7.02
    c. 2.31
    d. 1.93
    e. none of the above
    answer: a

20. One way to define the concentration of a buffer is to add the concentrations of components of the conjugate acid-base pair that produces the buffer. If 25.0 mL of a 0.25 M potassium monohydrogen phosphate was combined with 175 mL of 0.033 M potassium dihydrogen phosphate, what is the concentration of the resulting buffer? Assume the volumes to be additive.
    a. 0.029 M
    b. 0.031 M
    c. 12.025 M
    d. 0.060 M
    e. none of the above
    answer: b

21. Buffers are typically named by naming the non-hydrogen cation followed by the name of the most anionic form of the conjugate base. If solutions of ammonium dihydrogen citrate and ammonium hydroxide were combined, provide the name for the resulting buffer.
    a. citric acid buffer
    b. potassium citrate
    c. ammonium dihydrogen citrate
    d. ammonium citrate
    e. ammonium monohydrogen citrate
    answer: d

22. Which is true for an X-H:Y hydrogen bond in biological systems?
    a. X and Y typically are any combination of nitrogen and/or oxygen atoms
    b. It is the strongest when the axis of the orbital containing the lone pair and the axis of the X-H sigma bond meet at an angle less than 90°
    c. It is the strongest when the axis of the orbital containing the lone pair and the axis of the X-H sigma bond are co-linear
    d. a and b
    e. a and c
    answer: e

23. Icebergs are typically formed from fresh water (snow, etc.) which fell on land and contributed to glaciers before reaching ocean water. If ice has a density 0.917 g/mL, and ocean water has a density of 1.026 g/mL what percent of the iceberg the Titanic hit was actually above water?
    a. 2.5 %
    b. 10.6 %
    c. 97.5%
    d. 8.43 %
    e. 89.4%
    answer: b

24. Which of the following is true?
    a. Liquid and solid water have the same number of hydrogen bonds per molecule of water
    b. Liquid water has more hydrogen bonds per molecule than does solid water
    c. Solid water has more hydrogen bonds per molecule than does liquid water
    d. Hydrogen bonding is a major factor in the structure of gaseous water
    e. None is true
    answer: c

25. A single water molecule can participate in up to ___ hydrogen bonds.
    a. One
    b. Two
    c. Three
    d. Four
    e. Five
    answer: d

26. What impact does dilution have on a buffer?

    answer: The pH of a buffer is determined by the mole ratio of the components in an acid/base conjugate pair and the pKa involving those two components. The major impact of dilution is to reduce the buffering capacity of a specific buffer volume.

27. What is a pKa?

    answer: A pKa is the negative log (base 10) of an acid dissociation constant.

28. What is the ion product of water?

    answer: $[H^+][OH^-] = 1 \times 10^{-14} = Kw$

29. Write the three part equilibrium that exists between aqueous carbon dioxide, carbonic acid, and the bicarbonate ion.

    answer: $CO_{2\,(aq)} \leftrightarrow H_2CO_{3\,(aq)} \leftrightarrow H^+_{\,(aq)} + HCO_3^{-1}{}_{(aq)}$

18

Chapter 4 The Building Blocks of Proteins: Amino acids, Peptides and Polypeptides

1. Identify the relationship between cysteine and cystine.
   a. carboxylic acid and amide
   b. amide and carboxylic acid
   c. thiol and disulfide
   d. disulfide and thiol
   e. both are aromatic
   answer: c

2. Which amino acid has an R-group pKa closest to physiological pH?
   a. lysine
   b. histidine
   c. glutamate
   d. arginine
   e. aspartate
   answer: b

3. Which amino acid has the greatest percent nitrogen by mass?
   a. lysine
   b. tryptophan
   c. asparagine
   d. arginine
   e. histidine
   answer: d

4. Indicate the number of chiral carbons in: 4-hydroxyproline, glycine, threonine, leucine and isoleucine.
   a) 1 ,0 ,2 ,2 , 1
   b) 2, 0, 2, 1, 2
   c) 2, 1, 2, 1, 2
   d) 1, 0, 1, 1, 2
   e) 1, 0, 2, 2, 2
   answer: b

5. Which of the following amino acids have only two pKa's in the pH 1 to 14 range: Trp, Tyr, Cys, Lys, Asn, Arg?
   a) Trp, Tyr, Asn
   b) Tyr, Lys, Asn
   c) Trp, Asn
   d) Trp, Cys
   e) Cys, Asn
   answer: c

6. Which amino acids absorb in the UV portion of the spectrum?
    a. Cys and Met
    b. Asp and Glu
    c. Asn and Gln
    d. Arg, His, Lys
    e. Phe, Tyr, Trp
    answer: e

7. Protein amino acids are often used as unknowns in analytical procedures. One unknown was found to have an isoelectric point between 7.6 and 7.7. Identify the amino acid.
    a. Asp
    b. Arg
    c. Glu
    d. His
    e. Lys
    answer: d

8. A protein amino acid was found to have an isoelectric point near 10.7, what was the amino acid?
    a. Asp
    b. Arg
    c. Glu
    d. His
    e. Lys
    answer: e

9. Why are histidyl residues frequently utilized when mechanisms are proposed for enzymatically catalyzed reactions?
    a. Because of the partial aromatic character of its side chain.
    b. Because of the cyclic nature of its side chain.
    c. Because of the heterocyclic nature of its side chain.
    d. Because it can function as a proton donor or acceptor near physiological pH.
    e. Because of the planar nature of its side chain.
    answer: d

10. Proline is unique among the amino acids because
    a. it is the only amino acid whose alpha carbon is not chiral.
    b. it exists naturally in two diastereomeric forms.
    c. its alpha amino group is a primary amine.
    d. its alpha amino group is a secondary amine.
    e. its alpha amino group is a tertiary amine.
    answer: d

11. After the total acid hydrolysis of a small peptide, qualitative analysis showed the presence of Glu. Which of the following conclusions is best?
   a. the hydrolysis was incomplete
   b. Glu was present in the peptide
   c. Gln was present in the peptide
   d. Glu, Gln or both were present in the peptide
   e. Pro and/or Arg were oxidized during the hydrolysis
   answer: d

12. Cysteine can be converted to cystine by
   a. hydrolysis
   b. oxidation
   c. reduction
   d. dehydration
   e. denaturation
   answer: b

13. A mixture of Ala, Arg, and Asp in a pH 5.5 buffer was placed on a cation exchange column (the column is negatively charged) and eluted with the same buffer. What is the order of elution from the column?
   a. Arg, Ala then Asp
   b. Arg, Asp then Ala
   c. Asp, Ala then Arg
   d. Asp, Arg then Asp
   e. Ala, Asp then Arg
   answer: c

14. Ninhydrin is used to (as)
   a. a denaturant
   b. convert cysteine to cystine
   c. convert cystine to cysteine
   d. hydrolyze peptide bonds
   e. form a colored addition product with amino acids
   answer: e

15. Sanger's reagent, 2,4-dinitrofluorobenzene, is used to
   a. identify the C-terminus of a peptide
   b. identify the N-terminus of a peptide
   c. oxidize disulfide bonds to sulfonates
   d. react with aromatic amino acid side chains
   e. react with thiol side chains
   answer: b

16. Edman's reagent, phenyl isothiocyanate, is used to
    a. identify the C-terminus of a peptide
    b. identify the N-terminus of a peptide
    c. oxidize disulfide bonds to sulfonates
    d. react with aromatic amino acid side chains
    e. react with thiol side chains
    answer: b

17. The advantage of the Edman's reagent over Sanger's reagent
    a. complete oxidation of all disulfides
    b. more complete denaturation
    c. the process can be repeated on the remaining peptide
    d. more complete hydrolysis
    e. there is no advantage
    answer: c

18. What is the charge on the peptide Glu-Lys-Leu-Ser-Cys-Arg at a pH of 6?
    a. +2
    b. +1
    c. 0
    d. -1
    e. -2
    answer: b

19. At which of the following pH's will the peptide Asp-Leu-Ser-Phe-Lys-Glu have a charge of
    -3?
    a. 12
    b. 9
    c. 6
    d. 3.51
    e. 1
    answer: a

20. A tripeptide was treated with trypsin and produced a single amino acid and a dipeptide that
    had a UV absorbance. The same tripeptide, when exposed to chymotrypsin, produced a
    dipeptide and a single amino acid that had a UV absorbance. If hydrolysis of the tripeptide
    produced Ala, Lys, and Tyr, what was the sequence of the amino acids in the tripeptide?
    a. Ala-Lys-Tyr
    b. Ala-Tyr-Lys
    c. Tyr-Ala-Lys
    d. Tyr-Lys-Ala
    e. Lys-Ala-Tyr
    answer: d

21. A peptide was found to have a molecular mass of about 650 and upon hydrolysis produced Ala, Cys, Lys, Phe, and Val in a 1:1:1:1:1 ratio. The peptide upon treatment with Sanger's reagent produced DNP-Cys and exposure to carboxypeptidase produced valine. Chymotrypsin treatment of the peptide produced a dipeptide that contained sulfur and has a UV absorbance, and a tripeptide. Exposure of the peptide to trypsin produced a dipeptide and a tripeptide. Deduce the sequence of the peptide.
    a. Val-Ala-Lys-Phe-Cys
    b. Cys-Lys-Phe-Ala-Val
    c. Cys-Ala-Lys-Phe-Val
    d. Cys-Phe-Lys-Ala-Val
    e. Val-Phe-Lys-Ala-Cys
    answer: d

22. Treatment of a methionine containing peptide with cyanogen bromide followed by acid hydrolysis results in the formation of what amino acid?
    a. hydroxyproline
    b. hydroxylysine
    c. homoserine
    d. phosphoserine
    e. cystine
    answer: c

23. Lysyllysyllysine (a tripeptide composed only of lysine) was treated with the Sanger reagent followed by acid hydrolysis. Indicate the structure and ratio of the product(s).

    answer: N_α, N_ε-diDNP-Lysine and N_ε-DNP-Lysine in a 1 to 2 mole ratio.

24. During the formation of homoserine from methionine in the fragmentation of peptides with cyanogen bromide treatment, what is the source of the alcohol oxygen in the homoserine?

    answer: The homoserine lactone ring is formed by an approach of a lone pair of electrons on the amide carbonyl oxygen on the γ-carbon of the methionyl residue during the elimination of CH_3SCN. During this latter process the amide bond following the methionine is converted into a C=N. The hydrolysis of this C=N linkage produces the carbonyl oxygen of the lactone. Subsequent acid hydrolysis ultimately produces the homoserine. The actual alcohol oxygen originated as the carbonyl oxygen on the methionyl residue.

25. What (and why) is the single most important advantage of the solid phase peptide synthesis procedure?

    answer: The major advantage of the solid phase process is that purification steps, separation procedures, and solvent changes become filtrations. Mechanically this is less complicated than other methodologies and lends itself to automation.

23

26. Consider the structures of proline and tryptophan at low pH. The secondary nitrogen on proline is protonated and is positively charged. Is the ring nitrogen of tryptophan protonated? Why or why not?

answer: The indol nitrogen is SP^2 and the lone pair of electrons on the nitrogen contributes to the aromaticity of the five member ring and is unavailable for protonation.

27. An occasional instructor has been heard saying, "trypsin hydrolyzes after lysine or arginine residues, while chymotrypsin hydrolyzes after aromatic residues." Explain what the instructor means by "after" in the previous statement.

answer: Proteins and peptides are written in the N to C, left to right direction. Because it is the same as the direction we write the word "after" implies the same directionality, i.e., after Arg refers to the amide that follows Arg.

1. A reagent commonly employed to denature proteins is
   a. 2,4-dinitrofluorobenzene
   b. urea
   c. 2-mercaptoethanol
   d. phenylisothiocyanate
   e. cyanogen bromide
   answer: b

2. A reagent commonly employed to cleave disulfide bonds in proteins is
   a. 2,4-dinitrofluorobenzene
   b. urea
   c. 2-mercaptoethanol
   d. phenylisothiocyanate
   e. cyanogen bromide
   answer: c

3. Which structure is unique to collagen?
   a. the alpha helix
   b. the double helix
   c. the triple helix
   d. the beta structure
   e. the beta barrel
   answer: c

4. Which amino acids are found in large amounts in collagen?
   a. Pro, Phe, and Gly
   b. Pro, Gly, and OH-Pro
   c. Lys, His, and Arg
   d. Tyr, Phe, and Trp
   e. OH-Pro, OH-Lys, and Lys
   answer: b

5. Which of the following statements is true?
   a. all fibrous proteins have quaternary structure
   b. all globular proteins have quaternary structure
   c. all proteins have quaternary structure
   d. the interactions between the eight helical regions is partially responsible for the quaternary structure of myoglobin
   e. only proteins with more than one subunit have quaternary structure
   answer: e

6. In a peptide bond, the alpha amino nitrogen of proline can serve
    a. as a hydrogen bond acceptor only
    b. as a hydrogen bond donor only
    c. as both a hydrogen bond donor and acceptor
    d. as neither a hydrogen bond donor or acceptor
    answer: a

7. Which is not true for the following proteins: collegen, silk, hemoglobin, myoglobin, glucagon.
    a. they all have primary structure
    b. they all have secondary structure
    c. they all have tertiary structure
    d. they all have quaternary structure
    answer: d

8. In a globular protein, a leucine residue would most likely be found
    a. participating in hydrogen bonds involving R-groups
    b. on the outside surface
    c. on the interior
    d. both on the interior and outside surface
    e. leucine residues are rarely found in globular proteins
    answer: c

9. The planar geometry of the peptide bond is due to
    a. van der Waals forces
    b. disulfide bonds
    c. hydrogen bonds
    d. resonance effects
    e. inductive effects
    answer: d

10. The alpha helix and the beta sheet have much in common. Which of the following do the alpha helix and beta sheet not have in common?
    a. both are regular structures
    b. both are secondary structures
    c. both rarely contain proline
    d. both are compact structures
    e. both have hydrogen bonding
    answer: d

11. Which of the following does not apply to fibrous proteins?
    a. they are water insoluble
    b. they assume spherical shapes
    c. they have structural bio-functions
    d. they have repeating structures
    e. they are generally linear or planar
    answer: b

12. Which of the following peptides would most likely be found in the interior of a cell membrane?
    a. Asp-Glu-Gln-Asp
    b. Lys-Arg-Val-Lys
    c. His-Glu-Ser-Asn
    d. Met-Val-Ile-Phe
    e. Ser-Met-Asp-Gln
    answer: d

13. In a site directed mutagenesis experiment a particular Leu residue in a protein was changed to various other residues. Which of the following changes would have the greatest impact on the structure of the protein.
    a. Val
    b. Asp
    c. Ala
    d. Ile
    e. Phe
    answer: b

14. Reduced ribonuclease contains 8 cysteines. How many possible ways could the disulfides reform?
    a. 105
    b. 64
    c. 40320
    d. 1
    e. 4096
    answer: a (7x5x3x1)

15. White and Anfinsen in their studies on the renaturation of ribonuclease provided evidence that the conformation of a protein resulted from its amino acid sequence. In their experiments what sequence of events led to renaturation of a function ribonuclease and provided the basis for their conclusion?
    a. 1) denaturation and reduction, 2) dialysis, 3) oxidation
    b. 1) denaturation and reduction, 2) oxidation, 3) dialysis
    c. 1) denaturation, 2) oxidation, 3) reduction, 4) dialysis
    d. 1) oxidation, 2) dialysis, 3) denaturation, 4) reduction
    e. 1) oxidation and denaturation, 2) dialysis, 3) reduction
    answer: a

16. In a peptide the planar _____ meet at _____.
    a. alpha carbon, amides
    b. amides, alpha carbons
    c. carbonyls, alpha nitrogens
    d. amides, carbonyl carbons
    e. carbonyls, alpha carbons
    answer: b

17. The Ramachandran plot indicates:
    a. The fraction of the amino acids in a protein involved in alpha and beta structures
    b. The likelihood a particular amino acid is found in an alpha, beta or beta bend structure
    c. The angular arrangement of alpha helices within a protein
    d. The permitted angles of the two amides that meet at the alpha carbon of an amino acid
    e. The relationship between the helix axis and the fiber axis of a coiled coil
    answer: d

18. Amino acid(s) that is (are) typically found in beta (hairpin) bends include:
    a. Ala
    b. Gly
    c. Pro
    d. Gly and Pro
    e. Gly and Ala
    answer: d

19. The number of hydrogen bonds per amino acid.
    a. greater in parallel beta sheet than antiparallel beta sheet
    b. greater in antiparallel beta sheet than parallel beta sheet
    c. the same number in parallel and antiparallel
    d. hydrogen bonds are not a factor in parallel and antiparallel structures
    answer: c

20. Which of the following is not true about the common α-helix?
    a. the structure is right handed
    b. the distance along the helix axis is 5.4 Å per turn
    c. there are an integer number of amino acids per turn
    d. the amino acid R-groups are on the outside of the helix
    e. it is stabilized by hydrogen bonding
    answer: c

21. Which is true regarding the orientation of the R-groups in beta structures?
    a. In the parallel beta sheet structures the amino acid R-groups are all on the same side of the sheet.
    b. In the antiparallel beta sheet structure the R-groups on alternate strands are on the same side of the sheet.
    c. In parallel and antiparallel sheets consecutive R-groups on each peptide strand alternate sides of the sheet.
    d. Because of the free rotation around the alpha carbon, the R-groups can seek the least crowded region and are not restricted to a specific side of the sheet.
    e. a and b
    answer: c

22. Which is true concerning connections between strands of beta structure?
    a. Hairpin bends are common with parallel sheet structures.
    b. Right handed crossovers are common in parallel sheets and left handed crossovers are common in antiparallel sheets.
    c. Left handed crossovers are common in parallel sheets and right handed crossovers are common in antiparallel sheets.
    d. In antiparallel sheets right handed crossovers are uncommon and left handed crossovers are common.
    e. None of the above.
    answer: e

23. Invariant amino acids (IAA) are ones that appear in the same position in a particular protein when that protein is isolated from different species. In cytochrome c about one fifth of the IAA are glycine. If each of the twenty amino acids made an equal contribution to the IAA then only five percent of the IAA would be glycine; provide an explanation.

    answer: Because glycine has only a hydrogen as an R-group it has the greatest range of "permitted angles" on the Ranachangran plot. These IAA positions are quite likely to be hairpin bends where nature takes advantage of glycine's flexability.

24. The strength of the hydrogen bonds in an alpha helix are known to be stronger than those between the R-groups of two serines. Provide an explanation.

    answer: Because of the resonance form, the amide oxygen and nitrogen have a significant negative and positive character, respectively. This increases the polarization of the N-H bond and increases the electrostatic contribution in hydrogen bonds involving amides.

25. Indicate the locations of the R-groups in an alpha helix.

    answer: The R-groups extend radially from the "surface" of the helix, somewhat like the bristles of a bottle brush.

Chapter 6 Functional Diversity of Proteins

1. Which of the following statements is not true?
   a. Myoglobin has a higher affinity for oxygen than hemoglobin.
   b. At an oxygen pressure of 100 mm hemoglobin and myoglobin have equal affinity for oxygen.
   c. Myoglobin consists of a single peptide chain.
   d. Hemoglobin is better able to deliver oxygen at venous oxygen pressures.
   e. Myoglobin effectively transports oxygen from the lung to individual tissues.
   answer: e

2. Which of the following does not describe hemoglobin?
   a. Hemoglobin is an oxygen transport protein.
   b. Hemoglobin is an allosteric protein.
   c. Hemoglobin has quaternary structure.
   d. Copper ions are important in hemoglobin's ability to bind oxygen.
   e. Hemoglobin is composed, in part, of an iron containing heme group.
   answer: d

3. Hemoglobin is:
   a. a tetramer of four beta subunits.
   b. a tetramer of two alpha and two beta subunits.
   c. a tetramer of two beta and two gamma subunits.
   d. a tetramer of four alpha subunits
   e. a dimer of an alpha and a beta subunit.
   answer: b

4. To bind oxygen, the iron in hemoglogin must have a
   a. plus two oxidation state.
   b. plus three oxidation state.
   c. plus one oxidation state.
   d. zero oxidation state.
   e. any oxidation state.
   answer: a

5. Which of the following statements describes the oxygen binding curve of hemoglobin?
   a. Each of the four oxygens bind with equal facility.
   b. The binding of the first oxygen molecule enhances the binding of the other three oxygen molecules.
   c. The binding of the first oxygen molecule makes the binding of the other three oxygen molecules more difficult.
   d. The binding of the first oxygen molecule has no effect on the binding of the remaining three oxygen molecules.
   e. Each successive oxygen bound makes the remaining sites less likely to bind oxygen.
   answer: b

6. Which of the following amino acid changes would have the greatest effect on hemoglobin's ability to bind bis-phosphoglycerate (BPG)?
   a. Lys 82 to Arg
   b. Lys 82 to Phe
   c. Lys 82 to Val
   d. Lys 82 to Glu
   e. Lys 82 to His
   answer: d

7. The amino termini of the beta chains of hemoglobin are involved in binding BPG. If these amino termini were converted to amides, how would this affect hemoglobin's ability to bind BPG?
   a. There would be no effect.
   b. The BPG would bind more strongly.
   c. The BPG would not bind at all.
   d. The BPG would bind less strongly.
   e. Two molecules of BPG would bind.
   answer: d

8. In sickle cell anemia, the glutamic acid at position 6 of the beta chain is replaced with valine. Which of the following amino acid changes at position 6 would cause a minimal change?
   a. Glutamic acid to aspartic acid
   b. Glutamic acid to alanine
   c. Glutamic acid to isoleucine
   d. Glutamic acid to phenylalanine
   e. Glutamic acid to leucine
   answer: a

9. Under which of the following conditions will hemoglobin bind less oxygen?
   a. The pH increases from 7.0 to 7.2
   b. The oxygen pressure increases from 500 mm to 100 mm
   c. The concentration of carbon dioxide increases
   d. The BPG is removed
   e. none of the above
   answer: c

10. As the pH decreases, the ability of hemoglobin to bind oxygen
    a. increases
    b. decreases
    c. stays the same
    d. drops to zero
    answer: b

11. As the concentration of carbon dioxide decreases, the ability of hemoglobin to bind oxygen
    a. increases
    b. decreases
    c. stays the same
    d. drops to zero
    answer: a

12. Carbon dioxide binds to which site of the hemoglobin molecule?
    a. The heme iron
    b. The C-terminus amino acid
    c. The N-terminus amino acid
    d. The epsilon-amino group of a lysine
    e. The amide nitrogen of an asparagine
    answer: c

13. Which of the following components of vertebrate skeletal muscle has the greatest mass?
    a. actin
    b. troponin
    c. tropomyosin
    d. myosin
    e. all are about the same
    answer: d

14. According to the sequential model for allosteric proteins, which of the following statements is true for hemoglobin?
    a. Each of the four subunits in hemoglobin changes one at a time from the low affinity state to the high affinity state.
    b. The alpha subunits, then the beta subunits change from the low affinity to the high affinity state.
    c. The beta subunits, then the alpha subunits change from the low affinity to the high affinity state.
    d. Each of the four subunits in the hemoglobin tetramer is either the low affinity state or the high affinity state.
    e. The alpha subunits have low affinity and the beta subunits have high affinity for oxygen.
    answer: a

15. What is the function of actin?
    a. a major component of the thick filaments
    b. a major component of the thin filaments
    c. a rod shaped protein that binds the length of the actin filaments
    d. is involved in regulating muscle contraction
    e. is the precursor of troponin
    answer: b

16. What is the function of myosin?
    a. a major component of the thick filaments
    b. a major component of the thin filaments
    c. a rod shaped protein that binds the length of the actin filaments
    d. is involved in regulating muscle contraction
    e. is the precursor of troponin
    answer: a

17. What is the function of tropomyosin?
    a. a major component of the thick filaments
    b. a major component of the thin filaments
    c. a rod shaped protein that binds the length of the actin filaments
    d. is involved in regulating muscle contraction
    e. is the precursor of troponin
    answer: c

18. What is the function of troponin?
    a. a major component of the thick filaments
    b. a major component of the thin filaments
    c. a rod shaped protein that binds the length of the actin filaments
    d. is involved in regulating muscle contraction
    e. is the precursor of tropomyosin
    answer: d

19. Which of the following species does not bind preferentially to deoxyhemoglobin?
    a. BPG
    b. hydrogen ions
    c. hydroxide ions
    d. bicarbonate ions
    answer: c

20. Which of the following changes will not diminish the ability of hemoglobin to bind oxygen?
    a. the binding of BPG
    b. an increase in the number of hydrogen ions
    c. the increase in the oxidation state of the iron from 2+ to 3+
    d. the conversion of the hemoglobin tetramer into its individual subunits
    answer: d

21. Which of the following functions does not apply to hemoglobin?
    a. the ability to store oxygen in the peripheral tissues
    b. the ability to transport oxygen from the lung to the peripheral tissues
    c. the ability to transport carbon dioxide from the peripheral tissues to the lung
    d. the ability to stabilize pH
    e. a and d
    answer: a

22. With respect to a relaxed muscle, which of the following is true?
    a. the number of actin-myosin cross-bridges is maximized and the muscle is shortened
    b. the number of actin-myosin cross bridges is maximized and the muscle is stretched
    c. the number of actin-myosin cross bridges is minimized and the muscle is shortened
    d. the number of actin-myosin cross bridges is minimized and the muscle is stretched
    e. none of the above
    answer: d

23. What reaction is facilitated by carbonic anhydrase?
    a. $H_2CO_3 \rightarrow H_2O_2 + CO$
    b. $H_2CO_3 \rightarrow H^+ + HCO_3^{1-}$
    c. $2 H_2CO_3 \rightarrow H_2O + HOC(O)OC(O)OH$
    d. $2 H_2CO_3 \rightarrow H_2O_2 + HO_2CCO_2H$
    e. $CO_2 + H_2O \rightarrow H_2CO_3$
    answer: e

24. The isoelectric point for normal human hemoglobin is 6.9. What statement can you make about the isoelectric point you would expect for hemoglobin S (sickle-cell)?

    answer: Because a glutamate has been replaced with a valine residue, an acidic amino acid has been replaced with one bearing no charge. The isoelectric point is expected to increase. The actual isoelectric point for hemoglobin S is near 7.1

25. Would you expect the Glu_6 that is changed to a valine in hemoglobin S, to be on the surface or interior of the protein?

    answer: Normally the majority of the hydrophylic amino acid residues are on the surface of a water soluble globular protein.

26. What impact would you expect the replacement of Glu_6 of hemoglobin with valine to have on the solubility of hemoglobin S?

    answer: The replacement of a hydrophilic amino acid with one that is hydrophobic is expected to impact the water solubility of that protein. Hemoglobin is less soluble in water than is hemoglobin. Because hemoglobin S is less soluble it is more likely to crystallize (form tubular fibers). These latter structures actually contribute to the sickling of the red blood cells.

27. Deoxyhemoglobin S is known to be less soluble than oxyhemoglobin S. How does this contribute towards an explanation of the fact that a sickle cell crisis can be brought on by a reduced oxygen partial pressure.

    answer: When exposed to a reduce oxygen partial pressure a greater portion of the hemoglobin molecules would be in the deoxy-form. If the solubility limits of the deoxy-form is exceeded then tubular fibers form which in turn cause the sickling of the red blood cells.

34

Chapter 7 Methods of Characterization and Purification of Proteins

1. A mixture of urease (pI = 5.1, mol. wt. 482,700), catalase (pI = 5.6, mol. wt. 247,500), lactoglobin (pI 5.2, mol. wt. 37,100) and hemoglobin (pI 6.9, mol. wt. 64,500) were applied to a gel-exclusion chromatography column. What was their order of elution?
   a. urease, lactoglobin, catalase, hemoglobin
   b. hemoglobin, catalase, lactoglobin, urease
   c. urease, catalase, hemoglobin, lactoglobin
   d. lactoglobin, hemoglobin, catalase, urease
   answer: c

2. A mixture of urease (pI = 5.1, mol. wt. 482,700), catalase (pI = 5.6, mol. wt. 247,500), lactoglobin (pI 5.2, mol. wt. 37,100) and hemoglobin (pI 6.9, mol. wt. 64,500) were applied in a pH 6.5 buffer to a DEAE-cellulose chromatography column and eluted with the same buffer. What was their order of elution?
   a. urease, lactoglobin, catalase, hemoglobin
   b. hemoglobin, catalase, lactoglobin, urease
   c. urease, catalase, hemoglobin, lactoglobin
   d. lactoglobin, hemoglobin, catalase, urease
   e. cannot be determined from the information given
   answer: b

3. A mixture of urease (pI = 5.1, mol. wt. 482,700), catalase (pI = 5.6, mol. wt. 247,500), lactoglobin (pI 5.2, mol. wt. 37,100) and hemoglobin (pI 6.9, mol. wt. 64,500) were applied in a pH 6.5 buffer to a CM-cellulose chromatography column and eluted with the same buffer. What was their order of elution?
   a. urease, lactoglobin, catalase, hemoglobin
   b. urease, lactoglobin, and catalase elute together followed by hemoglobin
   c. hemoglobin, followed by urease, catalase, and lactoglobin eluting together
   d. lactoglobin, hemoglobin, catalase, urease
   e. none of the above
   answer: b

4. If you wanted to separate hemoglobin from hemoglobin S, which of the following procedures would be your first choice of methods?
   a. salting out procedures
   b. differential centrifugation
   c. ion exchange chromatography
   d. gel-exclusion chromatography
   e. ultracentrifugation
   answer: c

5. If you wanted to purify a monomer form from a dimer form of the same protein what would be your first choice of methods? Assume there is no equilibrium between the monomer and dimer forms of this protein.
   a. salting out procedures
   b. differential centrifugation
   c. ion exchange chromatography
   d. gel-exclusion chromatography
   e. ultracentrifugation
   answer: d

6. The rate of migration of a protein in electrophoresis depends upon the
   a. time of electrophoresis
   b. electrical potential
   c. net charge on the protein
   d. a, b and c.
   e. b and c
   answer: e

7. What is the function of the "gel" in many electrophoresis procedures?
   a. serves as an electrical conductor
   b. contributes physically to the separation of proteins
   c. provides a mechanical support after electrophoresis
   d. b and c
   e. a and b
   answer: d

8. Sodium dodecyl sulfate gel electrophoresis is often used for
   a. protein molecular mass determinations
   b. protein purification
   c. determination of protein sedimentation constants
   d. all of the above
   e. a and b
   answer: a

9. Isoelectric focusing characterizes proteins based upon differences in
   a. molecular weight
   b. the ratio of its component acidic and basic amino acids
   c. the shape
   d. the number of subunits
   e. the fraction of its component amino acids that are hydrophobic
   answer: b

10. Sodium dodecyl sulfate (SDS) is used to
    a. cleave cystine residues
    b. form cystine residues
    c. act as a buffer
    d. cleave peptide bonds
    e. denature proteins
    answer: e

11. Mercaptoethanol is used to
    a. cleave cystine residues
    b. form cystine residues
    c. act as a buffer
    d. cleave peptide bonds
    e. denature proteins
    answer: a

12. EDTA is used to
    a. cleave cystine residues
    b. form cystine residues
    c. complex metal ions
    d. denature proteins
    e. c and d
    answer: c

13. During the purification of a protein, which of the following should increase as the purification process proceeds?
    a. the volume of the sample
    b. the specific activity
    c. the total amount of protein
    d. the total number of units of protein
    e. the percent yield
    answer: b

14. Which of the following would be used to salt out a protein
    a. ammonium sulfate
    b. SDS
    c. mercaptoethanol
    d. EDTA
    e. a KH_2PO_4/ K_2HPO_4 mixture
    answer: a

15. Which of the following techniques cannot give information about the size of a protein?
    a. amino acid composition
    b. SDS-electrophoresis
    c. ion-exchange chromatography
    d. ultracentrifugation
    e. gel-exclusion chromatography
    answer: c

16. In a protein purification scheme, which of the following techniques would not be used first?
    a. ammonium sulfate precipitation
    b. differential centrifugation
    c. affinity chromatography
    d. isoelectric precipitation
    e. streptomycin precipitation
    answer: d

17. Separation of proteins by SDS-electrophoresis takes advantage of differences in
    a. amino acid composition of proteins
    b. protein isoelectric points
    c. protein size
    d. protein shape
    e. protein solubility
    answer: c

18. Separation of proteins by gel-exclusion chromatography takes advantage of differences in protein
    a. isoelectric points
    b. solubility
    c. size
    d. charge
    e. differences in aromatic amino acid composition
    answer: c

19. Separation of proteins by ion exchange chromatography takes advantage of differences in protein
    a. isoelectric points
    b. solubility
    c. size
    d. charge
    e. differences in aromatic amino acid composition
    answer: a

20. Which of the following would allow you to determine the isoelectric point of a protein?
    a. protein quaternary structure
    b. protein solubility as a function of pH
    c. protein size
    d. protein shape
    e. protein tertiary structure
    answer: b

21. Which of the following protein purification techniques would not be a reasonable last step in a multi-step protein purification scheme?
    a. HPLC
    b. DEAE-chromatography
    c. CMC-chromatography
    d. affinity chromatography
    e. ammonium sulfate precipitation
    answer: e

22. Why is a protein typically the least soluble at its isoelectric point?

    answer: At its isoelectric point a protein has no net charge and no net repulsion between molecules of the same protein. At any pH besides the isoelectric point each molecule of a protein will have the same charge and there will be a net repulsion between individual molecules.

23. What is the difference in purpose in a stacking and a resolving gel in electrophoresis?

    answer: The stacking gel is before (above) the resolving gel. Stacking gels are less crosslinked and differ in pH from the resolving gels which allows more rapid migration of proteins through the stacking gel. The main reason for the stacking gel is to allow the focusing of the sample into a "plane" which then enters the resolving gel simultaneously. Within the resolving gel the proteins separate primarily on the basis of their electrophoretic mobility.

24. The pI for urease is 5.1 while that of hemoglobin is 6.9. What conclusions can you make about the amino acid composition of these two proteins?

    answer: The value for the isoelectric point reflects the ratio of [Arg] +[Lys] + [His] / [Asp] + [Glu], where the brackets reflect the "mole fraction" that the indicated amino acid residues contribute to the total protein. The greater the numerical value for this ratio the greater the numerical value for the pI.

25. What inherent problem would you encounter in the design of the ligand that would be attached to an insoluble polymer to make an affinity column? For example consider you are trying to purify an enzyme that has lysine as a substrate and you wish to attach lysine to a polymer (there are many activated polymers that are commercially available that will react with many different functional groups, so do not consider this to be the problem).

answer: Unless you know how the lysine binds to the enzyme, the polymer may block a major binding point between the lysine and the enzyme. Also, if the lysine is free to bind "completely" the enzyme may convert the polymer-substrate into a polymer-product which might not bind to the enzyme.

Chapter 8 Enzyme Kinetics

1. An apoenzyme is
    a. a coenzyme
    b. an enzyme with its cofactor
    c. an enzyme lacking its cofactor
    d. an allosteric enzyme
    e. a cofactor
    answer: c

2. An holoenzyme is
    a. a coenzyme
    b. an enzyme with its cofactor
    c. an enzyme lacking its cofactor
    d. an allosteric enzyme
    e. a cofactor
    answer: b

3. An enzyme affects the structure of which species?
    a. The structure of the substrate.
    b. The structure of the transition state.
    c. The structure of the product.
    d. The structure of the intermediate.
    e. The structure of the coenzyme
    answer: b

4. Which of the following affects the rate of a reaction?
    a. The number of reactant collisions per unit time.
    b. The energy of each collision.
    c. The orientation of the particles in each collision.
    d. The temperature.
    e. All of the above.
    answer: e

5. The rate of an enzyme catalyzed reaction
    a. increases as the temperature increases.
    b. decreases as the temperature increases.
    c. is not affected by changes in temperature.
    d. is maximized over a narrow range of temperatures.
    e. is a direct function of the Kelvin temperature
    answer: d

6. The rate of an enzyme catalyzed reaction
    a. is independent of pH over a broad range of pH values.
    b. is maximized over a narrow range of pH values.
    c. increases as pH increases
    d. decreases as pH increases
    e. is affected more by the $[OH^-]$ than $[H^+]$.
    answer: b

7. Which of the following is not true concerning competitive inhibition of an enzyme catalyzed reaction?
    a. The inhibition cannot be overcome by increasing the concentration of the reactant.
    b. The competitive inhibitor resembles the substrate.
    c. The competitive inhibitor binds to the same site as does the substrate.
    d. The competitive inhibitor does not allow the substrate to bind to the enzyme.
    e. The inhibition can be overcome by increasing the concentration of the reactant.
    answer: a

8. A Lineweaver-Burk (double reciprocal) analysis allows for the determination of
    a. V_{max} only
    b. K_m only
    c. V_{max} and K_m
    d. neither V_{max} nor K_m
    e. the reaction product
    answer: c

9. In which of the following types of inhibition does the inhibitor bind to the same site as the substrate?
    a. Non-competitive inhibition.
    b. Uncompetitive inhibition.
    c. Competitive inhibition.
    d. a and b.
    e. all of the above.
    answer: c

10. Which of the following statements is true?
    a. Competitive inhibition is reversible and can be overcome by increasing the concentration of the substrate.
    b. Competitive inhibition is irreversible and cannot be overcome by increasing the concentration of the substrate.
    c. Non-competitive inhibition is normally irreversible but can be overcome by increasing the concentration of the substrate.
    d. Competitive inhibition is reversible and cannot be overcome by increasing the concentration of the substrate.
    answer: a

11. Which of the following statements is true about non-competitive inhibition?
    a. K_m increases
    b. K_m decreases
    c. V_{max} increases
    d. V_{max} decreases
    e. V_{max} and K_m decrease
    answer: d

12. Which of the following statements is true about competitive inhibition?
    a. K_m increases
    b. K_m is unchanged
    c. V_{max} increases
    d. V_{max} is unchanged
    e. V_{max} and K_m are unchanged
    answer: a

13. In which of the following pathways does the enzyme exist in two different states?
    a. The ping-pong mechanism.
    b. The ordered pathway.
    c. The random pathway.
    d. The disordered pathway.
    answer: a

14. If $[S] = 2 K_M$, what is v?
    a. $v = V_{max}$
    b. $v = 2 V_{max}$
    c. $v = 0.5 V_{max}$
    d. $v = 0.67 V_{max}$
    e. $v = 1.5 V_{max}$
    answer: d

15. At what substrate concentration will $v = 0.1 V_{max}$?
    a. $[S] = 9 K_m$
    b. $[S] = 10 K_m$
    c. $[S] = 1/9 K_m$
    d. $[S] = 0.1 K_m$
    e. $[S] = 0.9 K_m$
    answer: c

16. The enzyme that catalyzes the following belongs to which enzyme class?

$$^-O_2CCH_2\ CH_2CO_2^- + FAD \rightarrow FADH_2 + \ ^-O_2CCH=CHCO_2^-$$

    a. oxidoreductase
    b. transferase
    c. hydrolase
    d. isomerase
    e. lyase
    answer: a

17. The enzyme that catalyzes the following belongs to which enzyme class?

$$^-O_2CCH=CHCO_2^- + H_2O \rightarrow$$
$$^-O_2CCH(OH)CH_2CO_2^-$$

    a. oxidoreductase
    b. transferase
    c. hydrolase
    d. isomerase
    e. lyase
    answer: e

18. The enzyme that catalyzes the following belongs to which enzyme class?

$$^-O_2CCH(OH)CH_2CO_2^- + NAD^+ \rightarrow NADH + H^+ + \ ^-O_2CC(O)CH_2CO_2^-$$

    a. oxidoreductase
    b. transferase
    c. hydrolase
    d. isomerase
    e. lyase
    answer: a

19. The enzyme that catalyzes the following belongs to which enzyme class?

$$^+H_3NCH(CH_3)CO2^- + \ ^-O_2CC(O)CH_2CH_2CO_2^- \rightarrow$$

$$^+H_3NCH(CH_2CH_2CO_2^-)CO_2^- + CH_3C(O)CO_2^-$$

    a. oxidoreductase
    b. transferase
    c. hydrolase
    d. isomerase
    e. lyase
    answer: b

20. The enzyme that catalyzes the following belongs to which enzyme class?

$$^-O_2CCH{=}CHCO_2^- \ + \ NH_3^+ \ \rightarrow \ ^+H_3NCH(CH_2CO_2^-)CO_2^-$$

    a. oxidoreductase
    b. transferase
    c. hydrolase
    d. isomerase
    e. lyase
    answer: e

21. During steady state conditions which of the following item(s) stay constant?
    a. [E]
    b. [ES]
    c. [P]
    d. [S]
    e. [E] and [ES]
    answer: e

22. During steady state conditions which of the following item(s) increases?
    a. [E]
    b. [ES]
    c. [P]
    d. [S]
    e. [E] and [ES]
    answer: c

23. During steady state conditions which of the following item(s) decreases?
    a. [E]
    b. [ES]
    c. [P]
    d. [S]
    e. [E] and [ES]
    answer: d

24. Indicate which is true about enzymes.
    a. Enzymes are permanently changed during the conversion of substrate into product.
    b. Enzymes interact irreversibly with their substrates.
    c. Enzymes change the energy difference between substrates and products.
    d. Enzymes reduce the energy of activation for the conversion of reactant into product.
    e. Enzymes increase the energy content of the products.
    answer: d

25. Kinetic parameters (K_m, V_{max}) for enzymes are determined:
    a. By measuring the rate of the enzymatic reaction as a function of the substrate concentration.
    b. By measuring the rate of the enzymatic reaction as a function of the enzyme concentration.
    c. By using enzyme inhibitors.
    d. By determining the amount of product formed as a function of the amount of enzyme.

    answer: a

26. The K_m is:
    a. The time for half of the substrate to be converted to product.
    b. The time for all of the substrate to be converted to product.
    c. The [S] that gives half of the maximum reaction rate.
    d. The [S] that gives the maximum reaction rate.
    e. The [P] that is produced when the enzyme is saturated with the substrate.

    answer: c

27. Ping pong enzymatic mechanisms typically occur:
    a. with all enzymes that have two substrates.
    b. with enzymes that have two substrates that bind simultaneously.
    c. with enzymes that have two substrates that bind sequentially.
    d. with enzymes that have two products.
    e. when a single substrate is turned into two products.

    answer: c

28. Comment on the typical reaction sequence of substrate binding and product formation with enzymes that show ping pong kinetics.

    answer: Typically substrate one binds, the nature of the enzyme is changed, product one leaves, substrate two binds, the enzyme returns to its original state, and product two leaves.

29. Explain why iodoacetamide often is an irreversible inhibitor of enzymes.

    answer: Iodoacetamide reacts with thiols (RSH) to produce $RSCH_2C(O)NH_2$ where R is the the majority of the protein. If a thiol(s) is critical for the enzymatic activity or the protein structure then the iodoacetamide inhibits the enzyme. Because the iodoacetamide is covalently added to the protein, the inhibition is irreversible.

30. Many animated portrayals of enzymes in action, e.g. ads for detergents that contain proteolytic enzymes, will have the enzyme relentlessly attacking the stain until it is destroyed. What is wrong with the ideas presented in these ads?

   answer: The basic problem is that many dramatizations of enzymes imply that enzymes have senses, know where their substrates are and can propel themselves towards the substrate. All of these ideas are wrong. The previously mentioned processes are diffusion controlled and enzymes lack intelligence, mobility and senses.

Chapter 9 Mechanisms of Enzyme Catalysis

1. An endopeptidase cleaves peptide bonds
    a. beginning with the N-terminal amino acid.
    b. beginning with the C-terminal amino acid.
    c. beginning with internal amino acids of a protein.
    d that follow sulfur containing amino acids.
    e. none of the above.
    answer: c

2. Which of the following enzymes is not a serine protease?
    a. Elastase
    b. Trypsin
    c. Chymotrypsin
    d. Carboxypeptidase
    answer: d

3. The enzyme trypsin preferentially catalyzes the cleavage of peptide bonds
    a. on the C-side of Lys and Arg.
    b. on the N-side of Lys and Arg.
    c. on the C-side of Tyr, Trp, and Phe.
    d. on the N-side of Tyr, Trp, and Phe.
    e. from the N-terminus of the protein.
    answer: a

4. The enzyme chymotrypsin preferentially catalyzes the cleavage of peptide bonds
    a. on the C-side of Lys and Arg.
    b. on the N-side of Lys and Arg.
    c. on the C-side of Tyr, Trp, and Phe.
    d. on the N-side of Tyr, Trp, and Phe.
    e. from the N-terminus of the protein.
    answer: c

5. Which model characterizes an enzyme's tendency to modify its shape in order to accommodate its substrate?
    a. The lock-and-key model.
    b. The modified substrate model.
    c. The transformed fit model.
    d. The induced fit model.
    e. None of the above.
    answer: d

6. Which of the following amino acids does not play an important role in the active site of serine proteases?
   a. Lysine
   b. Serine
   c. Histidine
   d. Aspartate
   e. None of the above
   answer: a

7. Serine proteases are irreversibly deactivated by treatment with diisopropylfluorophosphate. This observation shows the importance of
   a. aspartic acid in the active site.
   b. histidine in the active site.
   c. serine in the active site.
   d. alanine in the active site.
   e. the importance of this observation is not understood.
   answer: c

8. The enzyme ribonuclease cleaves
   a. P-O bonds on the 5′ carbon of ribose in RNA
   b. P-O bonds on the 3′ carbon of ribose in RNA
   c. peptide bonds
   d. the bases from RNA
   e. phosphodiester bonds in DNA
   answer: a

9. The enzyme RNaseA is inhibited by treatment with iodoacetate. This observation shows the importance of which amino acid side chain is in the catalytic site of RNaseA?
   a. Val
   b. Lys
   c. His
   d. Ser
   e. Arg
   answer: c

10. Treatment of RNaseA with fluorodinitrobenzene reduces or destroys the activity of this enzyme. This observation shows the importance of which amino acid side chain in the catalytic activity of RNaseA?
    a. Val
    b. Lys
    c. His
    d. Ser
    e. Arg
    answer: b

11. In order for RNaseA to cleave phosphodiester bonds, the 3′ and 5′ bases, respectively, must be
    a. a purine and any base
    b. any base and a purine
    c. any base and a pyrimidine
    d. a pyrimidine and any base
    e. any base and any base
    answer: d

12. The cyclic intermediate which forms in the RNaseA cleavage of phosphodiester bonds is a
    a. cyclic 3′,5′-phosphate diester.
    b. cyclic 2′,5′-phosphate diester.
    c. cyclic 2′,3′-phosphate diester.
    d. cyclic 2′,4′-phosphate diester.
    e. cyclic 4′,5′-phosphate diester.
    answer: c

13. Which of the following processes will not inhibit the catalytic activity of RNaseA?
    a. treating RNaseA with iodoacetate.
    b. converting the 5′-base from a purine to a pyrimidine.
    c. converting the 2′-OH group to a 2′-methyl ether.
    d. converting the 2′-OH group to a 2′-deoxy group.
    e. treating RNase with fluorodinitrobenzene
    answer: b

14. Which of the following statements best describes the catalytic activity of triose phosphate isomerase? It catalyzes the:
    a. interconversion of dihydroxyacetone phosphate and glyceraldehyde-3-phosphate.
    b. conversion of dihydroxyacetone phosphate to glyceraldehyde-3-phosphate.
    c. conversion of glyceraldehyde-3-phosphate to dihydroxyacetone phosphate.
    d. transfer of a phosphate from dihydroxyacetone phosphate to glyceraldehyde.
    e. transfer of a phosphate from glyceraldehyde-3-phosphate to dihydroxyacetone.
    answer: a

15. The intermediate that forms in the reaction catalyzed by triose phosphate isomerase is a
    a. cyclic 2′,3′-phosphate diester
    b. a cyclic enediolate
    c. an enediolate
    d. a thioester
    e. an enediol
    answer: c

16. The catalytic effect of an enzyme depends primarily upon the enzyme stabilizing the
    a. reactant(s)
    b. product(s)
    c. transition state
    d. binding of the reactant(s)
    e. none of the above
    answer: c

17. The important general base in triose phosphate isomerase is
    a. aspartate
    b. glutamate
    c. histidine
    d. lysine
    e. oxyanion form of serine
    answer: b

18. The intermediate that forms in the RNaseA catalyzed hydrolysis of phosphodiester bonds is a
    a. divalent intermediate.
    b. trivalent intermediate.
    c. tetravalent intermediate.
    d. pentavalent intermediate.
    e. none of the above
    answer: d

19. Which of the following do not provide common nucleophilic groups on enzymes that participate in covalent intermediates with substrates?
    a. aspartate
    b. serine
    c. cysteine
    d. lysine
    e. arginine
    answer: e

20. Which of the following is not one of the "broad themes used in discussing enzyme reaction mechanisms"?
    a. structural flexibility
    b. electrostatic effects
    c. general-acid and general-base
    d. hydrophobic effects
    e. nucleophilic or electrophilic catalysis by enzyme functional groups
    answer: d

21. What can typically be said about the rates of intermolecular vs. intramolecular reactions (assuming the same chemical reaction)?
    a. both occur at the same rate
    b. intermolecular reactions occur faster than intramolecular reactions
    c. intermolecular reactions occur slower than intramolecular reactions
    d. it depends only on the type of reaction considered
    e. none of the above
    answer: c

22. Serine proteases form a covalent intermediate with their substrates. This intermediate is
    a. an amide involving the carboxyl group of the serine.
    b. an amide involving the amino group of the serine.
    c. an ester involving the alcohol group of the serine.
    d. an ester involving the carboxyl group of the serine.
    e. none of the above.
    answer: c

23. The substrate specificity is determined by which enzyme bound amino acids in chymotrypsin and trypsin, respectively?
    a. alanine, lysine
    b. aspartate, lysine
    c. alanine, lysine
    d. serine, aspartate
    e. arginine, aspartate
    answer: d

24. Some non-serine proteases contain
    a. Fe^{3+}
    b. Mg^{2+}
    c. Co^{2+}
    d. Mn^{2+}
    e. Zn^{2+}
    answer: e

25. *p*-nitrophenol acetate is:
    a. a reagent used to acetylate lysyl side chains of enzymes
    b. a reagent used to convert thiols to disulfides
    c. reacts specifically with histidyl side chains
    d. reacts specifically with arginyl side chains
    e. a synthetic substrate for chymotrypsin
    answer: e

26. What is meant by induced fit?

    answer: Induced fit implies a shape change in an enzyme upon interacting with its substrate.

27. What comments can you make on the evolution of subtilisin, trypsin, chymotrypsin and pancreatic endopeptidases?

    answer: All are serine proteases with Asp, His, and Ser involved in the catalysis. Trypsin, chymotrypsin and pancreatic endopeptidases all have sufficiently homologous amino acid sequences and structural similarities to indicate divergent evolution from a common ancestor. Subtilisin has a sufficiently different amino acid sequence and structure to suggest a evolutionary origin distinct from the other proteases mentioned. However, the fact subtilisin uses the same catalytic triad as the other proteases listed, suggests a convergent evolution.

28. What is the major advantage of general-acid and general-base catalysis?

    answer: Extreme pH's are not needed if the electrophilic/nucleophilic nature of a group is increased by the addition/removal of a proton.

Chapter 10 Regulation of Enzyme Activities

1. An allosteric effector binds to
   a. the active site of an enzyme.
   b. a site other than the active site of an enzyme.
   c. both the active site and a site other than the active site.
   d. the substrate before it binds to the enzyme.
   e. the substrate after it binds to the enzyme.
   answer: b

2. Allosteric enzymes generally exhibit
   a. linear kinetics.
   b. hyperbolic kinetics.
   c. classical Michaelis-Menten kinetics.
   d. sigmoidal kinetics.
   answer: d

3. The binding of aspartate to aspartate transcarbamylase exhibits
   a. no cooperativity
   b. negative cooperativity
   c. positive cooperativity
   d. both negative and positive cooperativity
   answer: c

4. The experimental results of the treatment of aspartate transcarbamoyl tranferase with
p-hydroxymercuri-benzoate demonstrated that in this enzyme
   a. the catalytic and regulatory activities are exhibited by the same subunits.
   b. the catalytic and regulatory activities are exhibited by the different subunits.
   c. CTP binds only to the catalytic subunit.
   d. CTP binds only to the regulatory subunit.
   answer: b

5. Glycogen phosphorylase is an enzyme which catalyzes the conversion of
   a. glycogen to glucose.
   b. glycogen to glycogen phosphate.
   c. glycogen to glucose-1-phosphate.
   d. glycogen to glucose-6-phosphate.
   e. glycogen to fructose-6-phosphate.
   answer: c

6. Which of the following statements is true about the enzyme phosphorylase?
   a. Phosphorylase is phosphorylated on a threonine residue.
   b. Phosphorylase is phosphorylated on a serine residue.
   c. Phosphorylase is phosphorylated on a tyrosine residue.
   d. Phosphorylase is not phosphorylated.
   e. Phosphorylase is phosphorylated on a tryptophan residue.
   answer: b

7. Which of the following statements is true about the enzyme phosphorylase b?
   a. Phosphorylase b is phosphorylated on a threonine residue.
   b. Phosphorylase b is phosphorylated on a serine residue.
   c. Phosphorylase b is phosphorylated on a tyrosine residue.
   d. Phosphorylase b is not phosphorylated .
   e. Phosphorylase b is phosphorylated on a tryptophan residue.
   answer: d

8. Which of the following statements is true for allosteric enzymes?
   a. The substrate binds more tightly to the T form.
   b. The substrate does not bind to the T form.
   c. The binding of the substrate favors the transition of the T form to the R form.
   d. The binding of the substrate favors the transition of the R form to the T form.
   e. None of the above.
   answer: c

9. Which of the configurations for a tetrameric enzyme does the symmetry model allow?
   a. T T T T
   b. T R RT
   c. R T T R
   d. all of the above
   e. none of the above
   answer: a

10. Which of the following statements is true about allosteric enzymes?
    a. Allosteric inhibitors bind preferentially to the R form.
    b. Allosteric inhibitors bind preferentially to the T form.
    c. Allosteric inhibitors bind equally well to T and R forms.
    d. Allosteric inhibitors do not bind to the T form.
    e. None of the above.
    answer: b

11. Which of the following terms best describes the effect that CTP has on the enzyme aspartate carbamoyl transferase?
    a. Feedback inhibitor.
    b. Partial proteolysis.
    c. Zymogen activation.
    d. An enzyme cascade.
    answer: a

12. The protein thioredoxin is involved in which of the following processes?
    a. The reduction of cysteine residues.
    b. The phosphorylation of cysteine residues.
    c. The oxidation of cystine residues.
    d. The reduction of cystine residues.
    e. The deamination of cystine residues.
    answer: d

13. Pyridoxal phosphate is a prosthetic group in glycogen phosphorylase. It is covalently attached to a(n)
    a. arginine residue.
    b. histidine residue.
    c. lysine residue.
    d. serine residue.
    e. asparagine residue.
    answer: c

14. The important glycolytic enzyme, phosphofructokinase, is
    a. stimulated by ADP and inhibited by ATP.
    b. stimulated by both ADP and ATP.
    c. stimulated by AMP
    d. inhibited by both ADP and ATP
    e. stimulated by ATP and inhibited by ADP.
    answer: a.

15. The transition between the R conformation and the T conformation in allosteric enzymes involves changes in
    a. primary structure.
    b. secondary structure.
    c. tertiary structure.
    d. quaternary structure.
    e. covalent modification.
    answer: d

16. Under physiological conditions, which of the following processes is not an important method for regulating the activity of enzymes?
      a. Phosphorylation.
      b. Temperature changes.
      c. Adenyl addition.
      d. Disulfide reduction.
      e. Partial proteolysis.
      answer: b

17. Chymotrypsinogen to chymotrypsin can not be described as
      a. a partial proteolysis.
      b. an activation.
      c. a covalent modification.
      d. reversible.
      e. a control mechanism.
      answer: d

18. Which of the following is not a common reversible covalent modification method for controlling enzymes?
      a. phosphorylation.
      b. partial proteolysis of a precursor protein
      c. adenylation
      d. disulfide reduction.
      e. none of the above.
      answer: b

19. The amino acid residues typically modified in phosporylation/dephosphorylation regulatory processes are?
      a. Asp and Glu
      b. Gln and Asn
      c. Ser, Thr, and Tyr
      d. Cys
      e. Tyr, Phe, and Trp
      answer: c

20. The most common regulatory mechanisms for enzymes are?
      a. reversible covalent modification
      b. irreversible covalent modification
      c. allosteric control
      d. a and c
      e. a, b, and c
      answer: e

21. A zymogen is:
    a. an inactive precursor
    b. an active precursor
    c. an inactive product
    d. an active product
    e. none of the above
    answer: a

22. During the conversion of chymotrypsinogen to chymotrypsin, which of the following occurs?
    a. a peptide is cleaved from the C terminus of chymotrypsinogen.
    b. a peptide is cleaved from the N terminus of chymotrypsinogen.
    c. a peptide is cleaved from the middle of chymotrypsinogen.
    d. a phosphate moiety is added.
    e. a cystine is reduced.
    answer: b

23. When chymotrypsin is formed from chymotrypsinogen:
    a. there is a major conformational change in the shape of the protein.
    b. there is a major aligning of the catalytic triad.
    c. the substrate binding site is created.
    d. a and b
    e. b and c
    answer: c

24. Enzyme regulatory mechanisms involving phosphorylation use a protein kinase, while those mechanisms involving dephosphorylation use a phosphatase. To which of the following enzyme class does a phosphatase belong?
    a. oxidoreductase
    b. transferase
    c. hydrolase
    d. isomerase
    e. lyase
    answer: c

25. Enzyme regulatory mechanisms involving phosphorylation use a protein kinase, while those mechanisms involving dephosphorylation use a phosphatase. To which of the following enzyme class does a protein kinase belong?
    a. oxidoreductase
    b. transferase
    c. hydrolase
    d. isomerase
    e. lyase
    answer: b

26. Protein kinases associated with the phosphorylation of proteins are typically:
    a. AMP dependent kinases
    b. cAMP dependent kinases
    c. involved in blood clotting
    d. are associated with many allosteric enzymes
    e. none of the above
    answer: b

27. The formation of blood clots involves a series of components that are:
    a. allosteric enzymes
    b. enzymes regulated by (de)phosphorylation
    c. enzymes activated by partial proteolysis
    d. enzymes regulated by adenylation
    e. blood clotting is a non-enzymatic process
    answer: c

28. The blood coagulation cascade has on occasion been called an amplification sequence. Explain what is meant by amplification.

    answer: Several of the steps involve functional enzymes as products of enzymatic reactions. Because enzymes can process many substrate molecules sequentially, there is a sequential multiplication of the initial signal.

29. What reason(s) can you provide why nature makes many proteolytic enzymes as zymogens?

    answer: If proteolytic enzymes were synthesized in their active form they would destroy the producing organ (pancreas) and also could not be stored.

30. Safety razor users have long know the merits of styptic pencils to stem the blood flow from nicks. Biochemistry instructors who use safety razors can often be seen substituting black board chalk for styptic pencils. Explain.

    answer: Styptic pencils contain a calcium salt which contribute calcium ions for the initiation of the clotting process. Chalk (calcium carbonate) can also provide calcium ions.

Chapter 11 Vitamins and Coenzymes

1. What coenzyme would you expect to be associated with the following conversion?

   $$^-O_2CCH_2CH_2CO_2^- \rightarrow {}^-O_2CCH=CHCO_2^-$$

   a. FAD / $FADH_2$
   b. NAD^+ / $NADH + H^+$
   c. Pyridoxal phosphate (B_6)
   d. Biotin
   e. B_{12}
   answer: a

2. What coenzyme would you expect to be associated with the following reaction?

   $$^+H_3NCH(CH_3)CO2^- + {}^-O_2CC(O)CH_2CH_2CO_2^- \rightarrow$$

   $$^+H_3NCH(CH_2CH_2CO_2^-)CO_2^- + CH_3C(O)CO_2^-$$

   a. FAD / $FADH_2$
   b. NAD^+ / $NADH + H^+$
   c. Pyridoxal phosphate (B_6)
   d. Biotin
   e. B_{12}
   answer: c

3. What coenzyme would you expect to be associated with the following conversion?

   $$^-O_2CCH(OH)CH_2CO_2^- \rightarrow {}^-O_2CC(O)CH_2CO_2^-$$

   a. FAD / $FADH_2$
   b. NAD^+ / $NADH + H^-$
   c. Pyridoxal phosphate (B_6)
   d. Biotin
   e. B_{12}
   answer: b

4. What coenzyme would you expect to be associated with the following conversion?

$$RCH_2CH_2COSCoA \rightarrow RCH=CHCOSCoA$$

a. FAD / FADH$_2$
b. NAD$^+$ / NADH + H$^+$
c. Pyridoxal phosphate (B$_6$)
d. Biotin
e. B$_{12}$
answer: a

5. What coenzyme would you expect to be associated with the following conversion?

$$^+H_3NCH(CH_2CH_2CH_2CH_2NH_3^+)CO_2^- \rightarrow {}^+H_3N(CH_2)_5NH3^+ + CO_2$$

a. FAD / FADH$_2$
b. NAD$^+$ / NADH + H$^+$
c. Pyridoxal phosphate (B$_6$)
d. Biotin
e. B$_{12}$
answer: c

6. What coenzyme would you expect to be associated with the following conversion?

$$^-O_2CCH(CH_3)COSCoA \rightarrow {}^-O_2CCH_2CH_2COSCoA$$

a. FAD / FADH$_2$
b. NAD$^+$ / NADH + H$^+$
c. Pyridoxal phosphate (B$_6$)
d. Biotin
e. B$_{12}$
answer: e

7. What coenzyme would you expect to be associated with the following conversion?

$$CH_3COSCoA + CO_2 + ATP \rightarrow {}^-O_2CCH_2COSCoA + ADP + Pi$$

a. Pyridoxal phosphate (B$_6$)
b. Biotin
c. Ascorbic acid (vitamin C)
d. B$_{12}$
e. Lipoic acid
answer: b

8. Which of the following contain an AMP moiety?

$1 = NADH \quad 2 = Biotin \quad 3 = CoASH \quad 4 = FAD \quad 5 = Lipoic\ acid$

   a. none of the combinations listed below
   b. 1, 4, and 5
   c. 1, 3, and 4
   d. 1, 2, and 3
   e. 1, 2, 3, 4, and 5
   answer: c

9. Which of the following vitamins is involved in the hydroxylation of proline in collagen?
   a. Vitamin A
   b. Vitamin B_1
   c. Vitamin B_6
   d. Vitamin C
   e. Vitamin D
   answer: d

10. The pyruvate-thiamine adduct observed in the action of thiamine pyrophosphate cannot be converted to
   a. ammonia
   b. acetoin
   c. acetate
   d. acetaldehyde
   e. acetylphosphate
   answer: a

11. Which of the following processes does not describe the action of NAD^+?
   a. It is an oxidizing agent.
   b. It is a hydride ion acceptor.
   c. It is a reducing agent.
   d. It is an electron acceptor.
   e. It is a water soluble substrate.
   anwser: c

12. Which of the following enzyme types would not use pyridoxal-5′-phosphate as a coenzyme?
   a. an aldolase
   b. an alcohol dehydrogenase
   c. a transaminase
   d. a racemase
   e. a decarboxylase
   answer: b

13. An important intermediate in enzyme catalyzed reactions that use pyridoxal-5′-phosphate as a coenzyme is a
    a. Schiff base
    b. semiquinone
    c. thioester
    d. thiol
    e. disulfide
    answer: a

14. Coenzyme A contains a reactive
    a. alcohol group.
    b. amino group
    c. carboxyl group
    d. thiol
    e. disulfide
    answer: d

15. Biotin uses which of the following reagents as a carboxylating agent?
    a. carbonate
    b. bicarbonate
    c. carbonic acid
    d. carbon dioxide
    e. carbonium ions
    answer: b

16. Vitamin B_{12} contains which of the following metals?
    a. magnesium
    b. copper
    c. cobalt
    d. chromium
    e. zinc
    answer: c

17. Vitamin K is involved in the synthesis of
    a. gamma carboxy glutamic acid
    b. beta carboxy glutamic acid
    c. alpha carboxy glutamic acid
    d. glutamic acid
    e. glutamine
    answer: a

18. Vitamin E is believed to be involved in the prevention of
    a. the reduction of polyunsaturated acids.
    b. the peroxidation of polyunsaturated acids.
    c. the peroxidation of glutamic acid.
    d. the peroxidation of aspartic acid.
    e. the oxidation of disulfides
    answer: b

19. The vitamin associated with vision is
    a. K
    b. D_3
    c. C
    d. A
    e. B_{12}
    answer: d

20. The vitamin which is associated with calcium and phosphate metabolism is
    a. K
    b. D_3
    c. C
    d. A
    e. B_{12}
    answer: b

21. Vitamin B_{12} is involved in which of the following types of reactions?
    a. racemization
    b. oxidation
    c. rearrangements on adjacent carbons
    d. reductions
    e. decarboxylations
    answer: c

22. Which of the following terms best describes the important functional group found in lipoic acid?
    a. a cyclic disulfide
    b. a disulfide
    c. a thiol
    d. a thioester
    e. an aldehyde
    answer: a

23. In enzymes which use lipoic acid as a coenzyme, to which amino acid side chain is the lipoic acid attached?
    a. serine
    b. cysteine
    c. lysine
    d. histidine
    e. tyrosine
    answer: c

24. Ascorbic acid is a(n)
    a. carboxylating agent
    b. decarboxylating agent
    c. oxidizing agent
    d. reducing agent
    e. acylating agent
    answer: d

25. An NAD^+ dependent epimerase involves which of the following reactions on the substance undergoing epimerization?
    a. a reduction followed by an oxidation
    b. an oxidation followed by a reduction
    c. two oxidations
    d. two reductions
    e. a carboxylation followed by a decarboxylation
    answer: b

26. One-carbon transfers are mediated by
    a. folate coenzymes
    b. coenzyme A
    c. lipoic acid
    d. thiamine pyrophosphate
    e. B_{12}
    answer: a

27. What necessitated nature to resort to the use of coenzymes?

    answer: Only a limited number of chemical reactions can be performed by the collection of functional groups available on the amino acids. The use of coenzymes allowed nature to go beyond these limitations.

28. Without resorting to chemical structures, how would you describe the function of pyridoxal phosphate to an organic chemist?

    answer: Pyridoxal phosphate facilitates, via resonance stabilization, the formation of anions on the alpha or beta carbon of an amino acid.

65

29. What general statement, in addition to their solubility, can you make on the biochemistry of the water-soluble versus the lipid soluble vitamins?

   answer: In general more is understood (mechanism) of the biochemical functions of the water-soluble vitamins.

Chapter 12 Metabolic Strategies

1. For which of the following type of lifestyles might a biochemist argue that sunlight does not provide the original source of energy?
    a. chemoautotroph
    b. heterotroph
    c. photoautotroph
    d. photoheterotroph
    answer: a

2. Most biochemistry students exhibit which type of metabolic lifestyle?
    a. chemoautotroph
    b. heterotroph
    c. photoautotroph
    d. photoheterotroph
    e. none of the above
    answer: b

3. Which of the following transitions might best apply to plants during sunrise each morning?
    a. photoautotroph to photoheterotroph
    b. dead to alive
    c. red to green
    d. heterotroph to photoheterotroph
    e. photoheterotroph to heterotroph
    answer: d

4. ATP and NADPH are metabolically important
    a. macromolecules
    b. primary couplers of anabolism and catabolism
    c. for deciphering the genetic code
    d. for binding related enzymes to membranes
    e. as starting points for metabolic pathways
    answer: b

5. The conversion of glucose to pyruvate is a multistep process requiring ten enzymes. If a mutant cell lacks one of these enzymes, which of the following happens?
    a. the cell will undergo reverse photosynthesis and give off protons
    b. the concentration of pyruvate will increase
    c. the cell will produce more of the other nine enzymes to maintain steady state
    d. the concentration of the metabolic intermediate which is the product of the missing enzyme will decrease
    answer: d

6. For a metabolic pathway involving multiple enzymes, the state in which each intermediate approaches chemical equilibrium with the preceding and following intermediates is called
    a. anabolic acidosis
    b. catabolic sufficiency
    c. steady state
    d. metabolic crossover
    e. stagnation
    answer: c

7. End product inhibition of multistep pathways is a major mechanism for cellular regulation. In this process, the
    a. product of the last enzyme inhibits the substrate of the last enzyme.
    b. product of the first enzyme inhibits the last enzyme in the pathway.
    c. first enzyme in the pathway is inhibited when ATP concentrations are depleted.
    d. product of the last enzyme inhibits the first enzyme in the pathway.
    e. end product causes an intermediate enzyme to form a different product and cause a branch in the pathway.
    answer: d

8. Regulated enzymes frequently exhibit a higher-order response to substrate concentration than non-regulated enzymes, which are commonly first-order. If enzyme "G" is first-order and enzyme "N" is fourth-order,
    a. enzyme N will be more sensitive to changes in substrate concentration than enzyme G.
    b. enzyme G is said to be cooperative.
    c. a graph of velocity versus substrate concentration for enzyme G will have the shape of a parabola.
    d. enzyme G will require energy from ATP to produce a product, but not enzyme N.
    answer: a

9. Which of the following is not true about the energy charge of a cell?
    a. it involves a ratio of ATP, ADP, and AMP
    b. it varies between values of 0 and 1
    c. catabolic reaction sequences tend to be inhibited as it increases
    d. it requires the enzyme adenylate kinase to be maintained
    answer: b

10. Which of the following is true for the analysis of a multistep metabolic pathway by the production of mutants?
   a. number of possible mutants = number of enzymes
   b. number of enzymes > number of genes
   c. number of complementation groups = number of enzymes
   d. cannot be used for anabolic pathways
   e number of complementation groups = number of enzymes minus one
   answer: c

11. Which of the following will probably not affect the rate at which an intermediate in a metabolic pathway is utilized by the appropriate enzyme?
   a. formation of a multienzyme complex involving the enzymes of the pathway
   b. increasing availability of the substrate for the first enzyme in the pathway
   c. a decrease in the concentration of the product of the last enzyme in the pathway
   d. addition of water to the organism's growth medium
   e. increase the temperature from 30° C to 37° C
   answer: d

12. Although the amount of ATP produced from the catabolism of glucose is large, the concentration of ATP in the cell remains small. Which of the following is the best explanation:
   a. ATP is a catalyst.
   b. biochemistry is not perfectly efficient and much energy is lost as heat.
   c. the energy put into ATP by catabolic reactions is transferred to anabolic reactions, regenerating ADP.
   d. The production of ATP is a futile cycle or pseudocycle.
   e. ATP is not particularly stable.
   answer: c

13. Which of the following enzymes is likely not to be regulated?
   a. the first enzyme in a pathway
   b. the enzyme catalyzing a futile cycle
   c. the first enzyme following a branch point in metabolism
   d. the last enzyme in a multienzyme pathway
   answer: d

14. Which best describes a pathway?
   a. several genes under the same control mechanism
   b. a sequence of chemical events used to describe an enzyme mechanism
   c. a sequence of steps taken by a compound to pass through a membrane
   d. a sequence of metabolites and the enzymes that interconvert these metabolites
   answer: d

15. Which of the following is not a known organization of functionally related enzymes
    a. components in a multiprotein complex
    b. proteins in solution in the same compartment
    c. as components on the same side of a membrane
    d. enzymes clustered into a tube through which the metabolites pass
    answer: d

16. Energy charge =
    a. [ATP] + [NADH]

    b. $\dfrac{[ATP] + [ADP]}{[ATP] + [ADP] + [AMP]}$

    c. $\dfrac{0.5\ [ATP] + [ADP]}{[ATP] + [ADP] + [AMP]}$

    d. $\dfrac{[ATP] + 0.5\ [ADP]}{[ATP] + [ADP] + [AMP]}$

    e. $\dfrac{2\ [ATP] + [ADP]}{[ATP] + [ADP] + [AMP]}$

    answer: d

17. Adenylate kinase catalyzes which of these reactions?
    a. $2\ AMP \rightarrow ADP$
    b. $AMP + ADP \rightarrow ATP$
    c. $ATP \rightarrow 3\ AMP$
    d. $ATP + ADP \rightarrow 5\ ADP$
    e. $AMP + ATP \rightarrow 2\ ADP$
    answer: e

18. Which of the following best describes a futile cycle?
    a. A common name for the citric acid cycle.
    b. A term used to describe the cyclic nature of the ADP moiety in the ATP to ADP to ATP etc.
    c. The combination of a kinase/phosphatase that results in the uncontrolled hydrolysis of ATP.
    d. A term used to describe NADP(H) cycling between catabolism and anabolism.
    e. None of the above.
    answer: c

19. One of the commonly heard attempts at "biochemical humor" is: "Turn out the lights and plants turn into animals." Explain what is meant by this statement.

answer: Plants are photoautotrophs when the sun is shining; without sunlight plants are heterotrophs which is the same life style as animals.

20. Describe the relationship between catabolism, metabolism and anabolism.

answer: Metabolism is a collective term referring to the sum of catabolism and anabolism. Anabolism is the collection of reactions involved in breaking down (often oxidation) of more complex molecules to simpler ones (often carbon dioxide and water) with the production of ATP and NAD(P)H. Catabolism, often called biosynthesis, refers to the conversion (often involving reduction) of simpler molecules to more complex molecules with the consumption of ATP and NADPH.

21. An occasional biochemist will annotate a reaction in which ATP is produced with a $. Explain the meaning of the $.

answer: ATP can be thought of as energy currency, produced by one reaction consumed by another.

22. Instructors of biochemistry have often been heard using the term Grand Central Station to describe pyruvate or acetyl-CoA. What aspect of these substances is being described by this term?

answer: All the trains, buses, and subways meet at Grand Central Station. The amphibolic nature or the involvement of these metabolites as the convergent point of many catabolic and anabolic pathways is being described.

23. Explain how the non-steady state concentration of intermediary metabolites can be used to determine which enzyme activity is lacking in a mutant.

answer: Relative to the concentration of intermediary metabolites in the non-mutant cell, the concentration of intermediates prior to the enzyme should be increased. Conversely, the concentration of intermediates following the lesion should be relatively decreased.

24. What chemical forces important in protein secondary, tertiary and quaternary structure are also important in the stable association of sequential enzymes in membranes?

answer: Hydrophobic forces would be important for interaction with the lipid component of the membrane. Electrostatic attractions from ionic pairing or salt bridges may be important for association of different proteins within the nonpolar environment of the membrane.

25 True or False? Most enzymes can catalyze a reaction in only one direction and not in the reverse direction.

answer: false

26. True or False? Inhibitors of specific enzymes can serve the same role as genetic blocks in the biochemical analysis of a pathway.

answer: true

27. True or False? The organization of functionally-related enzymes on membranes is only possible in eukaryotes, not prokaryotes.

answer: false

28. True or False? Only photoautotrophs use CO_2 as their source of carbon.

answer: false

29. True or False? Chemoautotrophs are the only organisms that use the oxidation of inorganic compounds as their source of energy to produce ATP.

answer: true

30. True or False? All metabolic classifications of organisms can make ATP from the energy associated with the oxidation of organic compounds.

answer: false

Chapter 13 Structures of Sugars and Energy-Storage Polysaccharides

1. Which of the following does not apply to D-glucose?
   - a. an aldohexose
   - b. a monosaccharide
   - c. an epimer of D-mannose
   - d. an isomer of D-galactose
   - e. can exist as either the α– or β-anomer

   answer: d

2. Conversion of an α–anomer to a β-anomer is termed
   - a. rotation
   - b. mutation
   - c. mutarotation
   - d. epirotation
   - e. epimutation

   answer: c

3. D-tagatose can be formed from D-galactose by the action of an isomerase. Which of the following is true about D-tagatose?
   - a. D-tagatose is an aldohexose
   - b. D-tagatose is the C-4 epimer of D-mannose
   - c. D-tagatose is the C-4 epimer of D-fructose
   - d. D-tagatose is the enantiomer of D-fructose
   - e. D-tagatose is an aldopentose

   answer: c

4. Without considering anomeric forms, deduce how many different D-aldopentoses are there.
   - a. 2
   - b. 4
   - c. 6
   - d. 8
   - e. none of the above

   answer: b

6. Without considering anomeric forms, deduce how many different D-ketopentoses are there.
   - a. 2
   - b. 4
   - c. 6
   - d. 8

   answer: a

7. If α-D-galactose has a specific rotation of 144.0°, and the β-form has a specific rotation of 52.0°, what percent of the galactose is α-D-galactose if the equilibrium mixture has a specific rotation of 80.5°?

    a. 70%
    b. 60%
    c. 40%
    d. 30%
    e. none of the above
    answer: d

8. The equilibrium mixture of fructose contains 36.9 % α-form and 63.1 % β-form. The α-D-fructose has a specific rotation of -21.0° and the β-form has a specific rotation of -133.5°. What is the specific rotation of the equilibrium mixture?

    a. + 8.0°
    b. - 8.0°
    c. +92.0°
    d. -92.0°
    e. none of the above
    answer: d

9. How many different compounds can be described as two D-glucoses linked (1,4)? Consider configurations but not conformations when answering this question.

    a. 1
    b. 2
    c. 3
    d. 4
    e. none of the above
    answer: d

10. A β-D-aldopentofuranose has how many chiral carbons?

    a. 0
    b. 2
    c. 3
    d. 4
    e. 5
    answer: d

11. An α-D-ketohexopyranose has how many chiral carbons?

    a. 0
    b. 2
    c. 3
    d. 4
    e. 5
    answer: d

12. Sugar alcohols are reduced sugars and are often used as "non-sugar sweeteners." Sugar alcohols can be thought of as sugars that have had their carbonyl group reduced to an alcohol. When glucose is reduced it produces only the sugar alcohol sorbitol. If sucrose is hydrolyzed then reduced, two sugar alcohols (sorbitol and mannitol) are produced. Predict the mole ratio of sorbitol:mannitol that would result from this chemical treatment of sucrose.

    a. 1:1
    b. 1:2
    c. 2:1
    d. 3:1
    e. 4:1
    answer: d

13. The sugar fucose is significant because it is a component of the ABO (and other) antagen and it is the milk sugar in monotremes. Fucose is unusual because it is a 6-deoxy-L-aldohexopyranose. How many different α-6-deoxy-L-aldohexopyranoses are there?

    a. 2
    b. 4
    c. 6
    d. 8
    e. more than 8
    answer: d

14. The sugar fucose is significant because it is a component of the ABO (and other) antagen and it is the milk sugar in monotremes. What is "unusual" about an α-6-deoxy-L-aldohexopyranose?

    1 = the deoxy aspect   2 = the "L" aspect   3 = the aldo aspect   4 = the pyranose aspect

    a. 1 and 2
    b. 1 and 3
    c. 2 and 3
    d. 2 and 4
    e. 3 and 4
    answer: a

15. The terms reducing and non-reducing are often applied to the ends of polysaccharides. How would you describe the ratio of the reducing:nonreducing ends in glycogen and cellulose?

	Glycogen	Cellulose
a.	1:1	1:1
b.	1:1	many:1
c.	many:1	1:1
d.	1:many	1:1
e.	1:1	1:many

answer: d

16. The conversion of α-D-glucose to β-D-glucose involves
    a. an equilibrium
    b. an isomerization
    c. an epimerization
    d. a reduction
    e. an oxidation
    answer: a

17. The conversion of glucose to maltose
    a. an equilibrium
    b. a hydrolysis reaction
    c. an dehydration process
    d. an epimerization
    e. a mirror image
    answer: c

18. The conversion of α-D-glucopyranose to α-D-glucofuranose involves
    a. an equilibrium
    b. an hydrolysis reaction
    c. a dehydration process
    d. an epimerization
    e. a mirror image
    answer: a

19. The conversion of glucose to fructose involves
    a. an equilibrium
    b. an oxidation reaction
    c. an isomerization process
    d. a reduction process
    e. a mirror image
    answer: c

20. The conversion of starch to glucose involves
    a. an equilibrium
    b. an hydrolysis reaction
    c. a dehydration process
    d. an epimerization
    e. a mirror image
    answer: b

21. The conversion of glucose to galactose involves
    a. an equilibrium
    b. an hydrolysis reaction
    c. a dehydration process
    d. an epimerization
    e. a mirror image
    answer: d

22. The conversion of D-glucose to L-glucose involves
    a. an equilibrium
    b. an hydrolysis reaction
    c. a dehydration process
    d. an epimerization
    e. a mirror image
    answer: e

23. What is sucrose?
    a. A disaccharide composed of galactose and glucose
    b. A disaccharide composed of glucose
    c. A disaccharide composed of galactose and fructose
    d. A disaccharide composed of glucose and fructose
    e. A monosaccharide
    answer: d

24. What is cellobiose?
    a. A disaccharide composed of galactose and glucose
    b. A disaccharide composed of glucose
    c. A disaccharide composed of galactose and fructose
    d. A disaccharide composed of glucose and fructose
    e. A monosaccharide
    answer: b

25. What is maltose?
    a. A disaccharide composed of galactose and glucose
    b. A disaccharide composed of glucose
    c. A disaccharide composed of galactose and fructose
    d. A disaccharide composed of glucose and fructose
    e. A monosaccharide
    answer: b

26. What is lactose?
    a. A disaccharide composed of galactose and glucose
    b. A disaccharide composed of glucose
    c. A disaccharide composed of galactose and fructose
    d. A disaccharide composed of glucose and fructose
    e. A monosaccharide
    answer: a

26. β-Methylglucoside has how many bulky groups in axial positions?
    a. 0
    b. 1
    c. 2
    d. 3
    e. 4
    answer: a

27. α-Methylglucoside has how many bulky groups in axial positions?
    a. 0
    b. 1
    c. 2
    d. 3
    e. 4
    answer: b

28. Corn syrup, an important sweetener, is formed from corn starch by treatment with pressurized steam and an acid catalyst. What would you expect to make up the major portion of corn syrup?

    answer: The anomeric forms of glucose and maltose are the major components of corn syrup.

29. Sucrose crystallizes readily, while hydrolyzed sucrose defies crystallization. Provide a chemical explanation for this phenomenon.

    answer: Because sucrose is a nonreducing sugar there is only one form in solution (i.e., it is only one compound). After hydrolysis, the glucose and fructose produced each consist of an open chain, anomeric, pyranose, and furanose forms. This complex mixture does not have a single predominant molecular species, and does not crystallize readily

30. Iodine dissolved in ethyl alcohol produces a yellowish-brown solution while when it is dissolved in hexane it produces a blue solution. What does this tell you about the nature of the blue starch-iodine complex?

    answer: The carbon bound hydrogens in axial positions line the helix and the iodine finds itself in an alkane-like environment (vs. the hydrophilic hydroxy portions of the glucose residues).

31. Starch and cellulose both have a potential aldehyde group but are considered to be non-reducing. Explain.

    answer: "In general the chemistry of a polymer is the chemistry of its middle." Yes, there is an aldehyde but its significance is masked by the mass of the polymer.

Chapter 14 Glycolysis, Gluconeogenesis, and the Pentose Phosphate Pathway

1. Why are sugars converted from six-carbon units to three carbon units in glycolysis?
    a. They must be made smaller to fit inside the enzymes.
    b. To prevent formation of furanose and pyranose ring structures.
    c. Carbohydrates with an even number of carbons cannot be utilized by anabolic pathways.
    d. Because conversion of two carbon units would be a metabolic dead end and the cell would die.
    e. None of the above
    answer: e

2. Which of the following three-carbon molecules is at a different oxidation level from the other molecules?
    a. D-glyceraldehyde
    b. L-glyceraldehyde
    c. dihydroxyacetone phosphate
    d. glycerol
    e. all four molecules are at identical oxidation levels
    answer: d

3. Conversion of an α-anomer to a β-anomer is termed
    a. rotation
    b. mutation
    c. mutarotation
    d. epirotation
    e. epimutation
    answer: c

4. Which of the following is a true statement concerning either starch or glycogen?
    a. Starch is normally found in pigeon flight muscles.
    b. Starch contains primarily β(1,6) glucoside linkages.
    c. Glycogen is normally found in pigeon liver tissue.
    d. Glycogen contains primarily β(1,6) glucoside linkages.
    e. Glycogen and starch contain equimolar amounts of glucose and galactose.
    answer: c

5. Which of the following enzymes might be essential for providing an organism with the ability to digest wood?
    a. glycogen phosphorylase
    b. b-amylase
    c. a-amylase
    d. cellulase
    e. none of the above
    answer: d

6. Which of the following molecules is directly involved in an enzymatic step which results in oxidation?
    a. fructose-1,6-bisphosphate
    b. dihydroxyacetone phosphate
    c. glyceraldehyde-3-phosphate
    d. phosphoenol pyruvate
    e. glucose-6-phosphate
    answer: c

7. Which of the following is not an intermediate in glycolysis?
    a. glycerol-3-phosphate
    b. phosphoenol pyruvate
    c. glycerate-1,3-bisphosphate
    d. glycerate-2-phosphate
    e. glycerate-3-phosphate
    answer: a

8. Which of the following conversions is catalyzed by the enzyme 3-phosphoglycerate kinase?
    a. glycerate-2-phosphate to phosphoenol pyruvate
    b. glycerate-3-phosphate to glycerate-2-phosphate
    c. glycerate-1,3-bisphosphate to glycerate-3-phosphate
    d. glyceraldehyde-3-phosphate to dihydroxyacetone phosphate
    answer: c

9. Identify which of the following enzymes is not involved in the phosphorylation of an intermediate during glycolysis
    a. hexokinase
    b. phosphofructokinase
    c. 3-phosphoglyceraldehyde dehydrogenase
    d. pyruvate kinase
    answer: c

10. Which of the following represents the products of a reaction catalyzed by the enzyme hexokinase?
   a. D-glucose-6-phosphate and ADP
   b. D-glucose-1-phosphate and NADPH
   c. D-glucose-1-phosphate and ATP
   d. D-glucose-1,6-bisphosphate and AMP
   e. pyruvate and ATP
   answer: a

11. Animal liver contains an enzyme termed glucokinase which catalyzes the same reaction as hexokinase. One difference between these enzymes is that glucokinase has a Michaelis constant for glucose 1,000 times higher than that of hexokinase. What does this imply about the relationship between these two enzymes?
   a. Glucokinase is normally active and hexokinase is used when glucose levels are unusually high.
   b. Glucokinase normally catalyzes the reaction in the reverse direction from hexokinase.
   c. Glucokinase is present to protect the organism from mutations in the hexokinase gene.
   d. Glucokinase is not normally utilized for glycolysis.
   e. None of the above
   answer: d

12. During the glycolytic conversion of D-glucose to pyruvate, how many of the carbon atoms in the original hexose are lost as CO_2
   a. none
   b. one
   c. two
   d. three
   e. six
   answer: a

13. Glycolysis of glucose could be described as the following sequence of events: first phosphorylation, second phosphorylation, cleavage and oxidation. The first ATP produced by glycolysis could be said to occur
   a. between the first and second phosphorylation
   b. between the cleavage and the oxidation
   c. during the cleavage
   d. during the oxidation
   e. after the oxidation
   answer: e

14. Which of the following as a starting material for glycolysis will result in the highest net yield of ATP?
   a. glucose
   b. fructose
   c. glycogen
   d. they would all produce equivalent net amounts of ATP
   answer: c

15. The reaction catalyzed by lactate dehydrogenase involves
   a. transfer of electrons from lactate to pyruvate
   b. oxidation of pyruvate to lactate
   c. pyruvate and ATP as substrates
   d. reduction of lactate to lactose
   e. regeneration of NAD^+ required for the continuation of glycolysis
   answer: e

16. The standard free energy change of the glycolytic reaction catalyzed by which of the following enzymes is strongly positive because an increase in the number of molecules increases the entropy of the system?
   a. hexokinase
   b. phosphohexose isomerase
   c. phosphofructokinase
   d. aldolase
   e. triosephosphate isomerase
   answer: d

17. Which of the following enzymes catalyzes reactions in both glycolysis and gluconeogenesis?
   a. phosphofructokinase
   b. fructose bisphosphate phosphatase
   c. pyruvate kinase
   d. hexokinase
   e. none of the above
   answer: e

18. The use of nucleotides such as GTP in a metabolic reaction is energetically equivalent to the use of ATP. This is because the GDP that is produced is rephosphorylated to GTP at the expense of ATP by which of the following enzymes?
   a. adenylate kinase
   b. ATPase
   c. GTPase
   d. nucleoside diphosphate kinase
   e. phosphorylase a
   answer: d

19. The molecular intermediate between glucose-1-phosphate and starch (a glucose polymer) is which of the following?
   a. glucose-6-phosphate
   b. glucose-1,6-bisphosphate
   c. maltose
   d. ADP-glucose
   e. ATP-glucose
   answer: d

20. Which of the following changes in cellular metabolite concentrations might indicate a need to speed up the net rate of glycolysis?
   a. increasing ATP and citrate
   b. increasing ADP and AMP
   c. decreasing glucose and AMP
   d. increasing citrate and pyruvate
   e. increasing galactose from lactose
   answer: b

21. To obtain maximal cellular production of glucose, a cell would
   a. stimulate glycogen phosphorylase and fructose bisphosphatase
   b. inhibit glycogen phosphorylase and stimulate phosphofructokinase
   c. stimulate phosphofructokinase and inhibit fructose bisphosphase
   d. inhibit phosphofructokinase and glycogen synthase
   e. answers b and c
   answer: e

22. The hormones glucagon and epinephrine signal liver cells to breakdown glycogen and produce glucose. This hormone signal outside the cell is transduced into an intracellular signal of increased cyclic-AMP concentration. What are the enzymes which produce and then degrade cyclic-AMP, respectively?
   a. adenylate kinase and adenylate phosphatase
   b. adenylate cyclase and phosphodiesterase
   c. ATPase and adenylate kinase
   d. glycogen phosphorylase and glycogen synthase
   e. phosphorylase a and phosphorylase b
   answer: b

23. All but one of the following statements about the pentose phosphate pathway is correct. Which statement is false?
   a. It involves the production of CO_2 without production of ethanol.
   b. It is a major source of NADPH for anabolic pathways.
   c. It starts with the same product of the hexokinase-catalyzed reaction as the glycolytic pathway.
   d. It uses the same aldolase enzyme as the glycolytic pathway.
   e. It is a major source of carbohydrates with five carbons.
   answer: d

24. Pretend that a crazy biochemist has taken over a first grade classroom. She insists that the students add up the number of carbons in the biochemical molecules. What is the answer to the following problem: Glucose + Fructose + Dihydroxyacetone =
   a. ribose + ribose + erythrose
   b. glyceraldehyde + glycerol + xyulose
   c. ribose + sedoheptulose + glycerol
   d. sedoheptulose + sedoheptulose
   e. maltose + galactose
   answer: c

25. Many bacteria have a metabolic pathway for the catabolism of D-galactose (the 4-carbon epimer of D-glucose) which involves the production of D-tagatose-1,6-bisphosphate (the 4-carbon epimer of D-fructose-1,6-bisphosphate). The next step in the pathway is catalyzed by an aldolase which produces two three-carbon phosphorylated intermediates. What are these products?
   a. glycerate-3-phosphate and glyceraldehyde-3-phosphate
   b. dihydroxyacetone phosphate and D-glyceraldehyde-3-phosphate
   c. dihydroxyacetone phosphate and L-glyceraldehyde-3-phosphate
   d. glycerate-2-phosphate and dihydroxyacetone
   e. glycerate-3-phosphate and dihydroxyacetone
   answer: b

26. Which of the following factors is not important in the coordination of cellular metabolic pathways?
   a. the standard free energy change for individual reactions
   b. energy charge
   c. hormone messengers
   d. allosteric behavior of enzymes
   answer: a

27. How is hexokinase able to convert a reactant that exists at a micromolar concentration into a product with a concentration in the millimolar range?
   a. by the use of an induced fit mechanism
   b. by the use of an allosteric enzyme
   c. by its capacity to uniquely recognize glucose
   d. by coupling the reaction to the conversion of an ATP to ADP
   e. by its capacity to dissociate from its product
   answer: d

28. The covalent linkage(s) between 3-phosphoglyceraldehyde dehydrogenase and its substrate is (are)?
   a. an ester
   b. a thioester
   c. a thiohemiacetal
   d. a Schiff base (imine)
   e. answers b and c
   answer: e

29. In the conversion of glycerate-3-phosphate to glycerate-2-phosphate, which is (are) true?
   a. The phosphate moves from the third to the second carbon via a cyclic intermediate.
   b. The phosphate in the 2-position is provided by ATP.
   c. The glycerate-2-phosphate produced has the phosphate that was on the previous molecule of glycerate-3-phosphate processed by the enzyme.
   d. There is a glycerate-2,3-bisphosphate intermediate.
   e. answers c and d.
   answer: XXXXX

30. Which of the following is not a function of the pentose phosphate pathway?
   a. provide NADPH
   b. provide ATP
   c. provide pentoses for nucleic acid biosynthesis
   d. provide means to metabolize dietary pentoses
   e. provide erythrose-4-phosphate for the synthesis of aromatic rings
   answer: b

31. Why isn't pyruvate kinase just reversed for the gluconeogenesis pathway?

   answer: The energy barrier for the reverse reaction is greater than can be overcome by an ATP. The common way nature uses to overcome this problem is to use two (or more) ATP's in sequential reactions involving a stable intermediate compound. In this particular case oxaloacetate serves as the intermediary compound.

Chapter 15 The Tricarboxylic Acid Cycle

1. Which of the following is not one of the metabolic roles of the citric acid cycle?
    a. to complete the oxidation of carbohydrates.
    b. to provide appropriate starting compounds for metabolic pathways.
    c. to provide reduced intermediates as electron donors for energy production by oxidative phosphorylation.
    d. to permit yeast which has produced ethanol under anaerobic conditions to utilize the ethanol as a source of carbon for growth when exposed to oxygen.
    e. to remove the metabolic waste products of glycolysis as CO_2.
    answer: e

2. Which of the following pairs of scientists received much of the credit for discovery of the citric acid cycle?
    a. Embden and Meyerhoff
    b. Szent-Gyorgyi and Krebs
    c. Michaelis and Menten
    d. Lineweaver and Burk
    e. Watson and Crick
    answer: b

3. The three carbon dicarboxylic acid malonic acid was used to elucidate the sequence of intermediates that make up the citric acid cycle.   Malonate acts by mimicking which of the following compounds?
    a. fumarate
    b. malate
    c. succinate
    d. oxaloacetate
    e. pyruvate
    answer: c

4. Two carbons from a molecule of pyruvate enter the citric acid cycle by the action of which two enzymes?
    a. pyruvate kinase and enolase
    b. pyruvate carboxylase and malate dehydrogenase
    c. pyruvate dehydrogenase and citrate synthetase
    d. pyruvate carboxylase and dehydrogenase
    e. none of the above
    answer: c

5. Three carbons of pyruvate can enter the citric acid cycle by way of
   a. pyruvate dehydrogenase
   b. pyruvate decarboxylase
   c. pyruvate carboxylase
   d. pyruvate kinase
   e. citrate synthetase
   answer: c

6. Identify the TCA cycle enzyme which utilizes citrate as a substrate.
   a. aconitase
   b. citrate synthase
   c. ATP-citrate lyase
   d. fumarase
   e. succinate thiokinase
   answer: a

7. Which of the following enzymes associated with the TCA cycle does not use NAD^+ as a cofactor?
   a. isocitrate dehydrogenase
   b. α-ketoglutarate dehydrogenase
   c. malate dehydrogenase
   d. succinate dehydrogenase
   e. pyruvate dehydrogenase
   answer: d

8. Which of the following enzymes does not have Coenzyme A (CoASH) as a product of its reaction?
   a. aconitase
   b. citrate synthase
   c. malate synthase
   d. succinate thiokinase
   answer: a

9. Which of the following is the product of the fumarase enzyme
   a. fumarate
   b. malate
   c. pyruvate
   d. isocitrate
   e. succinate
   answer: b

10. Which of the following molecules has an odd (i.e., not even) number of carbon atoms?
    a. citrate
    b. isocitrate
    c. fumarate
    d. α-ketoglutarate
    e. oxaloacetate
    answer: d

11. The product of which enzyme is the substrate for fumarase?
    a. citrate synthase
    b. pyruvate carboxylase
    c. pyruvate dehydrogenase
    d. α-ketoglutarate dehydrogenase
    e. succinate dehydrogenase
    answer: e

12. Which of the following is not a coenzyme for the pyruvate dehydrogenase complex?
    a. CoASH
    b. thiamine pyrophosphate
    c. ATP
    d. FAD
    e. lipoic acid
    answer: c

13. Which of the following enzymes does not produce CO_2 as a product of its catalytic reaction?
    a. isocitrate dehydrogenase
    b. succinate dehydrogenase
    c. pyruvate dehydrogenase
    d. α-ketoglutarate dehydrogenase
    answer: b

14. Which of the following intermediates is a substrate for the only membrane-associated enzymatic activity in the TCA cycle?
    a. citrate
    b. fumarate
    c. oxaloacetate
    d. pyruvate
    e. succinate
    answer: e

15. The TCA cycle involves a series of oxidations (-2e) and decarboxylations (CO_2). Which of the following best describes the **order** in which these oxidations and decarboxylations occur in the cycle?
    a. $-2e + CO_2$; $-2e$; $-2e + CO_2$; $-2e-$
    b. $-2e$; $-2e + CO_2$; $-2e$; $2e + CO_2$
    c. $-2e + CO_2$; $-2e$; $-2e$; $2e + CO_2$
    d. $-2e + CO_2$; $2e + CO_2$; $-2e$; $-2e$
    e. $-2e$; $-2e$; $-2e + CO_2$; $2e + CO_2$
    answer: d

16. Even though a molecule like citrate is not asymmetric, it is utilized by the machinery of the TCA cycle in an asymmetric manner. Enzymes often synthesize or utilize such molecules in a spereospecific manner. What is a reason for this?
    a. citrate is produced from D-glucose not L-glucose
    b. mitochondria are not perfect spheres and are thus asymmetric
    c. enzymes are composed of chiral components and are themselves asymmetric
    d. when energy, such as that in ATP, is expended chiral molecules can be formed
    e. the chirality of the resulting molecules depend upon which direction around the TCA cycle is taken by a metabolite
    answer: c

17. The ultimate acceptor of the electrons removed by oxidative decarboxylation in the TCA cycle is
    a. NAD^+
    b. $NADP^+$
    c. FAD
    d. ATP
    e. O_2
    answer: e

18. The reaction catalyzed by succinate dehydrogenase has a standard free energy ($\Delta G^{o'}$) of approximately 0.0 Kcal/mol. Which of the following is a correct implication based on this fact?
    a. the reaction only proceeds because the mitochondria produce heat
    b. the reaction only proceeds because of the releasing of CO_2 puts pressure on the cycle
    c. there is no reason to believe that the reaction would not proceed in a forward direction
    d. there is sufficient energy released to produce one mole of ATP
    e. because you are dividing by zero the equilibrium constant is infinity
    answer: c

90

19. The standard free energy ($\Delta G^{o'}$) for the reaction catalyzed by malate dehydrogenase is +7.1 Kcal /mol. The standard free energy for the TCA cycle is -7.7 Kcal/mol. What can be inferred about the net metabolic flux if the product of the malate dehydrogenase reaction is added to a mitrochondrial extract?

    a. malate dehydrogenase will catalyze the reversal of the TCA cycle

    b. the second enzyme will catalyze metabolic flux in the forward (normal) direction

    c. the cycle will flow both forward and backward

    d. the outcome will depend on the steady state concentration of the metabolic intermediates

answer: d

20. Which term best represents the central role of the TCA cycle as an end product of catabolic pathways and a source of starting materials for anabolic pathways?

    a. amphibolic

    b. metabolic

    c. buffer

    d. isobolic

    e. pluralistic

answer: a

21. The glyoxylate cycle is a modification of the TCA cycle brought about by the synthesis of two new enzymes. The glyoxylate cycle is found in the cotyledons of germinating plants which store oils and fats, rather than starch. As a food reserve for the young seedling. Thought the glyoxylate cycle fatty acids are converted to precursors of carbohydrates and amino acids. This process can be summarized by which of the following?

    a. oxaloacetate + acetyl-SCoA → citrate

    b. 2 acetyl-SCoA → succinate

    c. acetyl-SCoA + pyruvate → succinate

    d. isocitrate → glyoxylate + succinate

    e. acetyl-SCoA + glyoxylate → malate

answer: b

22. Which of the following is an enzyme of the glyoxylate cycle?

    a. isocitrate dehydrogenase

    b. pyruvate kinase

    c. isocitrate lyase

    d. pyruvate carboxlyase

    e. glucose-6-phosphate phosphatase

answer: c

23. Which pairs of metabolic regulatory motifs would make biochemical sense for pyruvate dehydrogenase?
    a. inhibited by acetyl-SCoA and AMP
    b. inhibited by pyruvate and stimulated by ATP
    c. inhibited by ATP and AMP
    d. stimulated by ADP and NAD^+
    e. stimulated by ATP and inhibited by AMP
    answer: d

24. Which of the following would not be expected to stimulate the metabolic flux through the TCA cycle?
    a. conversion from anaerobic to aerobic
    b. increased demand for anabolic reactions
    c. decreasing energy charge
    d. increasing ratio of $NADH/NAD^+$
    e. release of the hormone epinephrine into the blood stream
    answer: d

25. In eukaryotic cells, significant biosynthesis occurs in the cytoplasm and requires mitochondrial metabolites such as NAD(P)H and acetyl-SCoA. How are these necessary biochemicals obtained by cytoplasmic enzymes?
    a. special proteins transport them out of the mitochondria
    b. mitochondria are really hollow and in equilibrium with the cytosol
    c. metabolites such as malate shuttle between the mitochondrion and cytosol, carrying reducing equivalents and carbon skeletons for biosynthesis
    d. the required components are synthesized in the cytoplasm
    e. none of the above
    answer: c

26. Which of the following organisms would not contain mitochondria capable of carrying out the TCA cycle reactions?
    a. plants
    b. fish
    c. yeast without oxygen
    d. bacteria
    e. fungi
    answer: c

27. How many chiral carbons are there in the eight chemical intermediates in the TCA cycle (do not consider the CoASH, FAD, NADH, and GTP and related components)?

    a. 0

    b. 1

    c. 2

    d. 3

    e. more than 3

    answer: d

28. In order for the citric acid cycle to completely oxidize a substance such as α-ketoglutarate, enzymes in addition to those in the citric acid cycle are needed. Which of the following is not needed for this process?

    a. pyruvate kinase

    b. pyruvate dehydrogenase

    c. pyruvate carboxylase

    d. phosphoenolpyruvate carboxykinase

    answer: c

29. Only a limited number of metabolites are transported across the mitochondrial membrane. Which of the following collections of metabolites associated with the TCA cycle are transported between the cytosol and the mitochondria?

    a. pyruvate, succinate, fumarate

    b. pyruvate, citrate, malate

    c. citrate, succinate, malate

    d. citrate, oxaloacetate, α-ketoglutarate

    e. α-ketoglutarate, succinate, oxaloacetate

    answer: b

30. In his original proposal Krebs included *cis*-aconitate in the citric acid cycle. This substance is no longer included in the cycle. Can you provide an explanation why?

    answer: *Cis*-Aconitate is an intermediate in the reaction catalyzed by aconitase (even providing the enzyme's name) but is not released by the enzyme. When cis-aconitate is supplied to aconitase it eventually is converted into isocitrate, which explains Kreb's observation and his inclusion of it in the cycle.

31. Which of the TCA cycle enzymes are regulated by what factors?

    answer: Pyruvate dehydrogenase is inhibited by high concentrations of ATP, NADH, and acetyl-SCoA. Citrate synthetase is inhibited by high concentrations of ATP and NADH. Isocitrate dehydrogenase is inhibited by ample amounts of NADH and inhanced by high levels of ADP and NAD^+. α-Ketoglutarate dehydrogenase is inhibited by high concentrations of NADH.

Chapter 16 Electron Transport, Proton Translocation, and Oxidative Phosphorylation

1. If you understand glycolysis, you should be able to reason out the answer to this question. Louis Pasteur made an observation over a century ago that is now called the Pasteur Effect. He measured the carbohydrate consumption rate when yeast, which was growing without oxygen, was placed in an oxygen containing environment. What do you expect Pasteur observed?
   a. carbohydrate consumption and cell growth increased
   b. carbohydrate consumption decreased and cell growth increased
   c. no change in carbohydrate consumption or cell growth rates
   d. carbohydrate consumption increased and cell growth decreased
   e. the yeast started to produce ethanol
   answer: b

2. Which of the following should be avoided if a cell is to extract the full amount of energy from each molecule of glucose which enters the glycolytic pathway?
   a. the reduced carbohydrate is oxidized to the level of CO_2
   b. NADH is converted to NAD^+ by lactate dehydrogenase
   c. mutations which decrease flow through the pentose phosphate pathway by 50%
   d. avoid environments that are below 30 °C
   e. have an aerobic life style
   answer: b

3. In the anatomy of the mitochondrion, the compartment inside the inner membrane is termed which of the following?
   a. matrix
   b. stroma
   c. intermembrane space
   d. intramembrane space
   e. cytoplasm
   answer: a

4. Which of the following biochemical lifestyles would probably avoid the need for mitochondria in the cell?
   a. chemoautotrophic
   b. photoautotrophic
   c. heterotrophic
   answer: a

5. The mitochondrion has a very small genome and is only able to produce relatively few of its own proteins. Consequently, most mitochondrial proteins are synthesized outside of the mitochondrion and transported into that organelle after synthesis. Once inside the mitochondrion, these enzymes are able to accomplish biosynthesis of most of the smaller molecules needed by mitochondrion. Which of the following components of oxidative phosphorylation are probably produced inside the mitochondrion?
   a. cytochrome c
   b. F_1
   c. F_o
   d. succinate dehydrogenase
   e. heme
   answer: e

6. If cytochrome c is denatured and the heme is removed from the apoprotein, which of the following is now absent?
   a. nitrogen
   b. oxygen
   c. electrons
   d. iron
   e. negative charges
   answer: d

7. Electron transfer by cytochromes is mediated by which of the following changes?
   a. $Fe^{1+} \rightarrow Fe^{2+}$
   b. $Fe^{2+} \rightarrow Fe^{3+}$
   c. $Fe^{3+} \rightarrow Fe^{4+}$
   d. $Fe^{2+} \rightarrow Fe^{4+}$
   e. $Cu^{2+} \rightarrow Cu^{1+}$
   answer: b

8. The final complex in the mitochondrial electron transport chain, which catalyzes the terminal transfer of electrons to molecular oxygen, is termed
   a. Complex III
   b. F_1/F_o
   c. succinate dehydrogenase
   d. cytochrome oxidase
   e. an iron-sulfur cluster
   answer: d

9. The mitochondrial inner membrane has a flavoprotein, glycerol-3-phosphate dehydrogenase, which oxidizes glycerol-3-phosphate. What is the product of this reaction?

    a. NADH + H$^+$
    b. dihydroxyacetone phosphate
    c. 3-phosphoglycerate
    d. glyoxylate
    e. pyruvate
    answer: b

10. Which of the following enzymes does not directly link the TCA cycle with the electron transport pathway of oxidative phosphorylation?

    a. malate dehydrogenase
    b. succinate dehydrogenase
    c. complex I
    d. complex II
    answer: a

11. Which of the following components of the electron transport is unlike the others in terms of the number of electrons transferred?

    a. cytochrome b
    b. FADH$_2$
    c. FMNH$_2$
    d. NADH
    e. ubiquinone
    answer: a

12. Cyanide is very toxic to most organism. The reason is that it interferes with the transfer of electrons to molecular oxygen near the end of the electron transport chain. If a suspension of actively respiring mitochondria are treated with cyanide, ubiquinone and cytochrome c will

    a. have unchanged redox states
    b. both become more oxidized
    c. both become more reduced
    d. ubiquinone will become more reduced and cytochrome c will become more oxidized
    e. ubiquinone will become more oxidized and cytochrome c will become more reduced
    answer: c

13. Atabrine or quinacrine are flavine analogs used medically in antimalarial suppressive therapy. These compounds have the ability to replace flavine-based cofactors in enzymes and thereby cause an inhibition of activity. Predict which of the following might happen if these compounds are added to an actively respiring suspension of mitochondria.

    a. they would have no effect
    b. they would increase the amount of ATP produced
    c. they would stimulate utilization of oxygen
    d. they would cause succinate concentrations to increase
    e. they would cause cytochrome c to become more reduced
    answer: d

14. Which of the following transfers electrons by diffusion on the outside of the membrane?

    a. ubiquinone
    b. complex II
    c. cytochrome a
    d. cytochrome b
    e. cytochrome c
    answer: e

15. Which of the following is the hydrophobic electron carrier which is mobile within the membrane?

    a. ubiquinone
    b. complex II
    c. cytochrome a
    d. cytochrome b
    e. cytochrome c
    answer: a

16. Given the following redox couples:

$$\text{succinate} + CO_2 + 2\,H^+ + 2\,e^- \rightarrow \alpha\text{-ketoglutarate} + H_2O \quad E^{o\prime} = -0.67\ V$$

$$\text{fumarate} + 2\,H^+ + 2\,e^- \rightarrow \text{succinate} \quad E^{o\prime} = -0.17\ V$$

If the solution is prepared with equilmolar concentrations of succinate, α-ketoglutarate, and fumarate, what would you expect to happen?

    a. α-ketoglutarate will increase in concentration
    b. succinate will increase in concentration
    c. fumarate will increase in concentration
    d. ATP will be formed
    e. none of the above
    answer: b

17. NADH interacts directly with which component of the electron transport chain?
    a. succinate dehydrogenase
    b. complex I
    c. complex II
    d. complex III
    e. cytrochrome oxidase
    answer: b

18. Succinate interacts directly with which of the following electron transport components?
    a. complex I
    b. NADH dehydrogenase
    c. complex II
    d. ubiquinone
    e. cytochrome oxidase
    answer: c

19. Electrons from complex I are transferred directly to which of the following electron transport components?
    a. complex Ia
    b. complex II
    c. complex III
    d. cytochrome c
    e. ubiquinone
    answer: e

20. Which of the following is actively involved in the transport of protons across the mitochondrial inner membrane?
    a. cytochrome c
    b. lactate dehydrogenase
    c. pyruvate dehydrogenase complex
    d. F_1
    e. ubiquinone
    answer: e

21. The mitochondrial respiration in some specialized plant tissues is insensitive to added cyanide. What would be a possible reason for this phenomenon?
    a. they transfer electrons to water
    b. they have a different terminal oxidase complex
    c. they are acidic and cyanide is converted to HCN gas
    d. they have an increased Km for cyanide
    e. they oxidize cyanide to thiocyanate
    answer: b

22. When supplied with succinate, a suspension of respiration-competent mitochondria would be expected to do which of the following?
    a. make the solution more basic
    b. pump protons into the matrix
    c. pump hydroxide ions into the solution
    d. pump cytochrome c into the solution
    e. increase water production upon addition of ADP
    answer: e

23. A suspension of mitochondria will make ATP upon addition of succinate at a pH of 7.0 in the presence of 0.1 mM histidine. If the conditions are identical, but the mitochondria are suspended in 50 mM sodium phosphate, ATP production is inhibited. What is a likely explanation?
    a. histidine acts as an uncoupler
    b. high histidine converts NADH to NADPH
    c. the high histidine concentration causes all the ADP in the mitochondrion to become immediately converted to ATP
    d. histidine acts as a strong buffer, preventing pH changes
    e. histidine is a detergent and causes the membranes to become leaky
    answer: d

24. Mitochondria can be "tricked" into producing ATP without electron donors such as succinate by which of the following?
    a. add sufficient acid to lower the solution pH from 7 to 6
    b. add sufficient base to raise the solution pH from 7 to 8
    c. add them to stomach acid (pH about 1)
    d. add both cyanide and rotenone
    e. add an uncoupler of oxidative phosphorylation
    answer: a

25. Which complex is directly responsible for ATP production?
    a. complex IV
    b. iron-sulfur complex
    c. F_o complex
    d. F_1 complex
    e. F_o/F_1 complex
    answer: e

26. Catabolism of glucose via the TCA cycle and oxidative phosphorylation produces how much excess energy relative to anaerobic organisms utilizing only glycolysis?
    a. twice as much
    b. ten to twenty times as much
    c. about thirty times as much
    d. over one hundred times as much
    e. no difference in energy, only a different end product(s)
    answer: b

27. Bacteria do not have mitrochondria. This implies which of the following?
    a. electron transport must pump protons out of the cytoplasm during oxidative phosphorylation
    b. protons from inside the cell must travel to the outside by means of the F_o channel to produce ATP
    c. bacteria do not have the enzymes of the TCA cycle
    d. bacteria are unable to utilize oxygen to enhance growth on glucose
    answer: a

28. Both hemoglobin and cytochrome c contain the heme moiety. Which of the following is true concerning these two proteins?
    a. the heme is covalently bound and undergoes a redox reaction in both roles
    b. the heme is covalently bound and undergoes a redox reaction only in hemoglobin
    c. the heme is covalently bound and undergoes a redox reaction only in the cytochrome c
    d. the heme is held by noncovalent forces and undergoes a redox reaction in the cytochrome c
    e. the heme is held by covalent forces and undergoes a redox reaction in hemoglobin
    answer: d

29. Identify the only mitochondrial electron-transport complex that does not contain (an) Fe-S cluster(s)
    a. complex I
    b. complex II
    c. complex III
    d. complex IV
    answer: d

30. Identify the only mitochondrial electron-transport complex that contains a bound FMN.
    a. complex I
    b. complex II
    c. complex III
    d. complex IV
    answer: a

31. Identify the mitochondrial electron-transport complex(es) that does(do) not contain (a) cytochrome(s).
    a. complex I
    b. complex II
    c. complex III
    d. complex IV
    e. a and b
    answer: a

32. Identify the only mitochondrial electron-transport complex that contains a bound FAD.
    a. complex I
    b. complex II
    c. complex III
    d. complex IV
    answer: b

33. Why on a mole ratio basis, is there much more ubiquinone than any of the mitochondrial electron transport complexes?

    answer: Ubiquinone is dissolved in the inner membrane and functions as an electron carrier from complex I or II to III. This carrier function is performed by diffusion within the membrane. Because diffusion is involved, ubiquinone is needed in a greater mole ratio.

34. Why do so many of the components of the electron transport and oxidative phosphorylation processes involve copper and iron?

    answer: These are the two most prevalent metals in biological systems that have two oxidation states. Because electron transport is a sequence of redox reactions it is "logical" that nature would have utilized these metals in the evolution of these systems.

101

Chapter 17 Photosynthesis and Other Processes Involving Light

1. The overall equation for photosynthesis in plants is a follows:

$$6 \, CO_2 + 6 \, H_2O + light \rightarrow C_6H_{12}O_6 + 6 \, O_2$$

This process is often referred to as the reverse of respiration in which carbohydrate is oxidized to CO_2 and water. The component from respiration which could replace light in the process above is
   a. red photons
   b. green photons
   c. blue photons
   d. acetyl-SCoA
   e. NADPH and ATP
   answer: e

2. It is estimated that the annual amount of carbon dioxide fixed each year is approximately 10^{11} metric tonnes (1 tonne = 1000 kg). It is also estimated that 10^9 tonnes of chlorophyll is produced on earth each year. Considering these estimates, which of the following statements about the number of carbon dioxide molecules fixed by each chlorophyll molecule is true?
   a. each chlorophyll fixes about 100 molecules of carbon dioxide
   b. since the molecular weights of carbon dioxide and chlorophyll are about 44 and 900 respectively, each chlorophyll fixes considerably more than 1000 molecules of carbon dioxide per year.
   c. because chlorophyll is synthesized from the fixed carbon dioxide, it is impossible to make any definite conclusions
   d. carbon dioxide is not fixed by individual chlorophyll molecules
   e. each chlorophyll fixes one carbon dioxide molecule and must then be resynthesized
   answer: d

3. It is estimated that the earth's atmosphere contains 7×10^{11} tonnes of carbon as carbon dioxide, with an additional 6×10^{11} tonnes of carbon as carbon dioxide in the ocean surface layers. What do these facts imply about the expected Km for carbon dioxide of the ribulose-1,5-bisphosphate carboxylase in land plants and in marine algae?
   a. there is not enough information for a conclusion
   b. the Km should be higher in land plants
   c. the Km should be lower in marine plants
   d. the Km should be lower in marine algae, but since the oceans do not contain O_2 there is no competition with carbon dioxide for the active site and no oxygenase reaction
   answer: a

4. Which of the following is not involved in the production of biochemical energy from light?
    a. chlorophyll
    b. ribulose-1,5-bisphosphate
    c. CF_1/Cf_o
    d. trans-membrane pH gradient
    e. Q cycle
    answer: b

5. The two processes of photosynthesis and respiration involve the interconversion of carbon dioxide and water to (or from) carbohydrate and oxygen. What is true about the equilibrium of these two processes?
    a. both processes are at equilibrium
    b. both processes have the same equilibrium
    c. photosynthesis is at equilibrium in the summer, but respiration is at equilibrium throughout the year
    d. since one process produces CO_2 and the other process utilizes CO_2, they have different equilibria
    answer: b

6. As a photon travels through a chloroplast and gives it energy to the process of photosynthesis, through which supramolecular structures will it sequentially pass?
    a. inner envelope, outer envelope, stoma, thylakoid membrane
    b. outer envelope membrane, inner envelope membrane, stroma, thylakoid membrane
    c. outer envelope membrane, inner envelope membrane, stroma, thylakoid membrane, thylakoid lumen
    d. outer envelope membrane, inner envelope membrane, thylakoid lumen, stroma, thylakoid membrane, stroma
    answer: b

7. A "grana" is
    a. primordial chloroplast in the megaspore mother cell
    b. the yellow chloroplast in leaf tissue germinated in total darkness
    c. a supramolecular array of chloroplasts which focus light like a lens
    d. folded stacks of thylakoid membranes
    e. an electron excited by a photon
    answer: d

103

8. When a suspension of thylakoid membranes is acidified to a pH less than 5, the amount of pheophytin is greatly increased. What might be the explanation?
    a. low pH hydrolyzes the phytol ester from the chlorophyll
    b. low pH causes the magnesium ion to be lost from chlorophyll
    c. low pH causes carotenoids to form dimers
    d. low pH causes the membranes to aggregate
    answer: b

9. The chlorophyll used in photosynthesis is found where in the chloroplast?
    a. inner envelope membrane
    b. outer envelope membrane
    c. stroma
    d. thylakoid lumen
    e. thylakoid membrane
    answer: e

10. Chlorophyll is able to absorb light of all but which of the following colors?
    a. blue
    b. green
    c. red
    d. white
    answer: b

11. The energy in a photon of light
    a. travels at 8 meters per second
    b. is greater as the wavelength increases
    c. is measured in units of "einsteins"
    d. decreases with increasing distance from the sun
    e. is a product of Planck's constant and the frequency
    answer: e

12. The energy in photons goes where when it is absorbed by a molecule?
    a. it is shared by the nucleus and the elecrons
    b. it is all absorbed by the neucleus
    c. it is partially absorbed by the nucleus, but the remainder results in the emission of a γ-ray
    d. it forces an electron to a higher energy orbital
    e. a portion of the energy excites the vibration of the molecule and the remainder is used to cause the photon to change direction (refraction)
    answer: d

13. To absorb the energy from a photon, which of the following must be true?
    a. the chlorophyll must be in a leaf
    b. the chlorophyll must be oxidized
    c. the chlorophyll must be reduced
    d. the chlorophyll's electrons must be in the ground state
    answer: d

14. If an antenna chlorophyll absorbs energy from a photon, which is the desired pathway for de-excitation?
    a. fluorescence
    b. resonance energy transfer
    c. donation of the excited electron
    d. return to the lower energy level with loss of energy as heat
    answer: b

15. The absorption of the energy in a photon has what effect on the standard redox potential of the molecule?
    a. it does not change because it is a constant
    b. the excited molecule has a higher redox potential
    c. the excited molecule has a lower redox potential
    answer: c

16. The collection of a reaction center with antenna pigment molecules is termed what?
    a. P700
    b. P680
    c. P870
    d. a photosystem
    e. a chlorophyll dimer
    answer: d

17. The electrons transferred in photosynthesis come from what molecule?
    a. water
    b. carbon dioxide
    c. sodium cyanide
    d. oxygen
    e. NADH
    answer: a

18. The electrons transferred in the light reactions of photosynthesis end up in which molecule?
   a. NADPH
   b. $FADH_2$
   c. water
   d. oxygen
   e. carbon monoxide
   answer: a

19. The experiments of Emerson and Arnold in the 1930's found that optimal rates of O_2 evolution required a period of about 20 milliseconds between flashes. That is because the slowest component in the photosynthetic electron transfer takes about that long to accept an electron, move through the membrane to the next component, and then transfer the electron. Which component is likely responsible for this delay?
   a. pheophytin
   b. P680
   c. P700
   d. plastoquinone
   e. cytochrome b_6f
   answer: d

20. Which of the following is not an antenna component of the higher plant photosystems?
   a. carotenoids
   b. plastocyanin
   c. chlorophyll a
   d. chlorophyll b
   answer: b

21. The electrons transferred in photosynthetic energy production
   a. pass sequentially through photosystem I and photosystem II.
   b. pass sequentially through photosystem II and photosystem I.
   c. are transferred from NADPH to water.
   d. drive the transport of protons from the thylakoid lumen to the stroma.
   e. are always represented by arrows pointing up.
   answer: b

22. Photosystem I and photosystem II
   a. absorb photons of the same wavelengths.
   b. must be excited by photons at equal rates for optimal efficiency.
   c. both produce oxygen.
   d. will continue to transfer electrons to each other if the temperature is dropped to absolute zero.
   e. occur in adjacent chloroplasts and are connected by electron carriers.
   answer: b

23. The energy from how many photons is required to produce one molecule of oxygen (O_2)?
    a. one
    b. two
    c. 3.14
    d. about 4
    e. about 8
    answer: e

24. When the chloroplast makes the transition from darkness to light, what would you predict about the pH of the stroma?
    a. it will increase for about five seconds and then decrease
    b. it will generally increase
    c. it will generally decrease
    d. it will oscillate around a fixed point (e.g., pH 7.42 ± 0.25
    answer: b

25. What term is used to describe the production of ATP during photosynthesis?
    a. oxidative phosphorylation
    b. photolysis
    c. photophobic
    d. photophosphorylation
    e. phosphorylation
    answer: d

26. If algae are incubated in the presence of light and $^{14}CO_2$ for one hour, which of the following compounds would likely contain the most radioactivity?
    a. starch
    b. ribulose-1,5-bisphosphate
    c. ribose-5-phosphate
    d. dihydroxyacetone phosphate
    e. glycerate-3-phosphate
    answer: a

27. Which of the following is not a product from the reaction of ribulose-1,5-bisphosphate and oxygen?
    a. 3-phosphoglycerate
    b. glycerate-3-phosphate
    c. 2-phosphoglycolate
    d. dihydroxyacetone phosphate
    answer: d

107

28. Why have some plants evolved the C-4 cycle in which CO_2 is first fixed into four-carbon organic acids?
    a. it results in an increase in CO_2 concentration in the vicinity of the active site of ribulose-1,5- bisphosphate carboxylase/oxygenase
    b. it allows growth under prolonged dark conditions such as the arctic winter
    c. it was selected by breeders because the extra acidity improves the flavor of many fruits
    d. it evolved because of the greenhouse effect and it provides an advantage for plants growing under increasing CO_2 levels
    answer: a

29. In the bundle sheath cells of C-4 plants malate ($^-O_2CC(OH)HCH_2CO_2^-$) is converted by a reaction involving decarboxylation and oxidation into pyruvate, What probably happens in the reaction?
    a. carbon-1 is lost as CO_2 then the alcohol is oxidized
    b. the alcohol is oxidized then carbon-1 is lost as CO_2
    c. carbon-4 is lost as CO_2 then the alcohol is oxidized
    d. the alcohol is oxidized then carbon-4 is lost as CO_2
    answer: d

30. In the conversion of pyruvate to phosphoenol pyruvate in mesophyll cells of C-4 plants the reaction is: pyruvate + ATP + Pi → phosphoenol pyruvate + PPi + AMP. What is the fate/source of the various phosphate moieties. Note: the phosphate moiety attached to the ribose is designated as α.
    a. α to AMP, β + Pi to PPi, γ to phosphoenolpyruvate
    b. α to AMP, β + γ to PPi, Pi to phosphoenolpyruvate
    c. Pi to AMP, β + γ to PPi, α to phosphoenolpyruvate
    d. Pi to AMP, α + β to PPi, γ to phosphoenolpyruvate
    answer: b

31. What is the metabolic source of the CO_2 released in the bundle sheath cells of C-4 plants?
    a. the decarboxylation of pyruvate
    b. half from each of the two carboxyl groups of malate
    c. the CO_2 that was added to phosphoenolpyruvate to make malate
    d. the carboxyl group of malate that came from phosphoenolpyruvate
    e. from the conversion of isocitrate to α-ketoglutarate
    answer: c

32. This chapter presented photophosphorylation, i.e., the light dependent formation of ATP from ADP. This represents the third fundamentally different means used by nature to produce ATP. Identify the two other methods for the formation of ATP.

    answer: Substrate level phosphorylation (e.g., 3-phosphoglycerate kinase), and oxidative phosphorylation in electron transport.

Chapter 18 Structures and Metabolism of Polysaccharides and Glycoproteins

1. Polysaccharides used primarily for the storage of metabolic energy contain which monosaccharides as constituents?
    a. only D-glucose
    b. only D-galactose
    c. both D-glucose and D-galactose
    d. N-acetylgalactosamine and N-acetylmeramic acid
    e. D-fructose
    answer: a

2. Which of the following enzyme activities is involved in the conversion of D-galactose to D-glucose?
    a. aldolase
    b. 2-epimerase
    c. 4-epimerase
    d. isomerase
    e. mutase
    answer: c

3. Which of the following describes the most abundant structural polysaccharide in nature?
    a. $\alpha(1,4)$ heteropolymer
    b. $\beta(1,4)$ heteropolymer
    c. $\alpha(1,4)$ homopolymer
    d. $\beta(1,4)$ homopolymer
    e. $\beta(1,6)$ homopolymer
    answer: d

4. Which of the following describes the most abundant energy storage polysaccharide in nature?
    a. $\alpha(1,4)$ heteropolymer
    b. $\beta(1,4)$ heteropolymer
    c. $\alpha(1,4)$ homopolymer
    d. $\beta(1,4)$ homopolymer
    e. $\beta(1,6)$ homopolymer
    answer: c

5. Pectins are important polysaccharide constituents of plant cells wall and are present in fruit juices. When fruit juices are boiled, mixed with sugar and cooled, they form gels. This is the basis for the preparation of jams and jellies. The molecular composition of the pectins causes the polymer chains to assume an extended conformation, which increases their sphere of influence and generates a significant viscosity in the surrounding medium. What could you surmise about the composition of pectins.
    a. They are composed of a large proportion of glucose residues
    b. They are composed of a large proportion of galactose residues
    c. They are relatively short chain polysaccharides
    d. They contain a relatively large number of negative charges.
    e. They contain a relatively large number of free radicals.
    answer: d

6. Which of the following is not caused by glycoproteins or glycolipids?
    a. clumping of red blood cells
    b. sickle cell anemia
    c. the human ABO blood group substances
    d. species specificity of egg fertilization
    answer: b

7. Which of the following does not contain N-glycoside linkages?
    a. glucosamine
    b. ATP
    c. glycoproteins in the *cis*-Golgi network
    d. glycoproteins in the *trans*-Golgi network
    answer: a

8. To which of the following residues would carbohydrates not be bound in glycoproteins?
    a. asparagine
    b. histidine
    c. hydroxylysine
    d. serine
    e. threonine
    answer: b

9. Which of the following enzymes would be required to add additional carbohydrates to a glycoprotein?
    a. aldolase
    b. kinase
    c. decarboxylase
    d. transferase
    e. phosphorylase
    answer: d

10. Which of the following is the primary sequence information which targets nascent polypeptide to the endoplasmic reticulum?
   a. CAA Box
   b. Pribnow Box
   c. proline elbow
   d. signal recognition particle (SRP)
   e. signal sequence
   answer: e

11. During transport of nascent glycoproteins into the endoplasmic reticulum, which cellular structures are anchored to the endoplasmic reticulum?
   a. chloroplasts
   b. lysosomes
   c. mitochondria
   d. nucleus
   e. ribosomes
   answer: e

12. Which of the following is unlikely to happen in the Golgi complex?
   a. addition of sugar residues to glycoproteins
   b. removal of sugar residues from glycoproteins
   c. glycosyl transfer
   d. oxidative phosphorylation
   e. post-translational modification of proteins
   answer: d

13. In which of the following organisms would formation of glycoproteins be unlikely to occur in the Golgi complex?
   a. bacteria
   b. humans
   c. pine trees
   d. mushrooms
   e. yeast
   answer: a

14. Dolichol phosphate serves what function in biosynthesis?
   a. it is an intermediate carrier of oligosaccharides destined for covalent attachment to proteins
   b. it transports nascent polypeptides across the endoplasmic reticulum
   c. it is a phosphate donor for activating glucose during glycolysis in the Golgi complex
   d. it is involved in the transfer of a phosphate form GTP to ADP, thereby producing ATP in the endoplasmic reticulum
   answer: a

15. Glycoproteins generally move from the medial-Golgi to the
    a. *cis*-Golgi
    b. *trans*-Golgi
    c. lysosome
    d. endoplasmic reticulum
    e. *meta*-Golgi
    answer: b

16. Dolichol phosphate and undecaprenol phosphate are what types of molecules
    a. carbohydrates
    b. lipids
    c. proteins
    d. antibiotics
    e. xenobiotics
    answer: b

17. Which of the following would be the most likely to diffuse across a biological membrane?
    a. glucosamine
    b. glucose
    c. UDP-glucose
    d. glucose-6-phosphate
    e. glucuronic acid
    answer: b

18. Which of the following is likely to be found containing the mannose-6-phosphate signal for which there is a glycoprotein acting as a Man-6-P receptor?
    a. hemoglobin
    b. hexokinase
    c. hydrolase
    d. myoglobin
    e. ribulose-1,5-bisphosphate carboxylase
    answer: c

19. Which of the following describes the role of dolichol phosphate in the glycosylation of O-linked oligosaccharides?
    a. it transfers the initial GalNAc to a serine or threonine
    b. it is necessary to recognize the sequence Asn-X-Ser(Thr)
    c. it is responsible for moving O-linked glycoproteins from the endoplasmic reticulum to the Golgi
    d. it enables glycosyltransferases to distinguish between a glycoprotein's N-terminus and C-terminus
    e. none of the above
    answer: e

20. The ABO blood type is based on two enzymes. The gene for one enzyme codes a glycosyltransferase that adds a terminal N-acetylgalactosamine residue to a core oligosaccharide, while the other gene codes for a similar gene that adds a galactose residue to the same site. People of O blood type are often called "universal donors." Which genotype would represent a person able to code for both enzymes?

    a. OO
    b. AB
    c. BB
    d. AO
    e. BO
    answer: b

21. Which of the following would have a potent biochemical effect in the endoplasmic reticulum?

    a. bacitracin
    b. penicillin
    c. tunicamycin
    d. vancomycin
    answer: c

22. Which of the following is a toxic plant protein able to bind to the surface of mammalian cells?

    a. lactam
    b. lecithin
    c. lectin
    d. lignin
    e. luciferin
    answer: c

23. Which of the following statements about bacterial cell walls is incorrect?

    a. they involve both carbohydrates and amino acids
    b. they are essential for survival of the bacteria
    c. because of crosslinks, each cell wall is composed of one gigantic molecule
    d. part of the synthesis takes place inside the cell and part outside the cell
    e. their composition was first elucidated by Louis Pasteur in 1884
    answer: e

24. Which of the following are the precursors of UDP-GlcNAc during the biosynthesis of the bacterial cell membrane?

    a. UTP + N-Acetylglucosamine-1-P
    b. UDP + N-Acetylglucosamine-1-P
    c. undecaprenol phosphate + N-Acetylglucosamine-1-P
    d. ubiquinone + N-Acetylglucosamine-1-P
    answer: c

25. Phosphonomycin is an antibiotic which relies on being an analog of phosphoenolpyruvate for activity. Which intermediate would be the product of the blocked reaction?
   a. UDP-GlcNAc
   b. UDP-GlcNAc-enolpyruvylether
   c. UDP-MurNAc
   d. UDP-MurNAc-Ala-Glu-Lys
   e. UDP-N-acetylmuramyl-pentapeptide
   answer: b

26. Cycloserine is an antibiotic which relies on being an analog of D-alanine for activity. Which intermediate would be the product of the blocked reaction?
   a. UDP-GlcNAc
   b. UDP-GlcNAc-enolpyruvylether
   c. UDP-MurNAc
   d. UDP-MurNAc-Ala-Glu-Lys
   e. UDP-N-acetylmuramyl-pentapeptide
   answer: e

27. Penicillin does which of the following?
   a. prevents cell wall breakdown in tomatoes
   b. inhibits the enzyme penicillinase
   c. prevents the transport of galactose into the bacterium
   d. prevents *Penicillium* mold from growing on bread
   e. prevents transpeptidation cross linking of bacterial cell walls
   answer: e

28. Which of the following is not a normal intermediate in the formation of oligosaccharides and polysaccharides from glucose?
   a. monophosphorylated hexose
   b. nucleoside diphosphate hexose
   c. N-acetylated hexosamine
   d. ketohexose
   answer: d

29. Which of the following is a heteropolymer?
   a. cellulose
   b. glycogen
   c. hyaluronic acid
   d. starch
   answer: c

30. Each glucose unit added to an oligosaccharide or polysaccharide increases the molecular weight by
    a. 150
    b. 162
    c. 180
    d. 198
    e. 212
    answer: b

31. Which of the following is true oligo- and polysaccharides are biosynthesized from monomers activated on the (A) end added to the (B) end of the polymer.

	A	B
a.	reducing	nonreducing
b.	reducing	reducing
c.	nonreducing	nonreducing
d.	nonreducing	reducing

    answer: a

32. In the biosynthesis of the peptide portion of the peptidoglycan, what is the source of the glycine?
    a. glycine
    b. glycyl-tRNA
    c. UDP-glycine
    d. glycyl-adenylate
    e. glycyl-phosphate
    answer: b

33. Which of the following is true concerning the peptide portion of peptidoglycans?
    a. contain only amide bonds analogous to those in proteins
    b. contain ester linkages
    c. contain amide bonds involving the epsilon amino group of lysine
    d. contain amide bonds involving the gamma carboxyl group of glutamate
    e. c and d
    answer: e

34. Cross linking in peptidoglycans involves
    a. linking of disaccharide components
    b. linking of monosaccharide components
    c. peptide bond formation with the release of a D-alanine
    d. use of lipid carriers
    answer: c

35. Explain why penicillin kills only growing bacteria.

answer: Penicillin inhibits the cross linking step in peptidoglycan biosynthesis. Peptidoglycan biosynthesis is only needed when bacteria are growing/dividing.

36. The most common amino sugars have the amino group on carbon 2. Provide a reason why.

answer: The most common amino sugars are aldoses and they are biosynthesized from the corresponding ketoses.

37. In a rash moment a famous biochemist (Fritz Lipmann) was heard to say there are more possible oligosaccharides than oligopeptides. What arguments can you make to support this statement?

answer: Even though there may be more common amino acids than common sugars, sugars can be connected together in many possible linkages, [α(1-6), β(1-6), α(1-4), β(1-4), α(1-3), β(1-3), α(1-2), β(1-2)], to say nothing about branched structures, pyranose or furanose forms.

1. Which of the following are normally unable to permeate a biological membrane?
   a. Na^+ and K^+
   b. glycerol
   c. CO_2
   answer: a

2. Which is probably a true statement about the relationship of a membrane to reseal quickly if
   breached, if its melting transition = Tm?
   a. The ability to reseal a membrane is greatest below the Tm
   b. The ability to reseal a membrane is greatest at the Tm
   c. The ability to reseal a membrane is greatest above the Tm
   answer: c

3. Which of the following is not found in membranes?
   a. proteins
   b. starch
   c. chlorophyll
   d. phospholipids
   e. glycolipids
   answer: b

4. The biological role of cholesterol is which of the following?
   a. to help reseal arteries which have developed leaks
   b. to make membranes more fluid
   c. to transfer light to P700 in photosynthesis
   d. to transfer electrons to chytochrome c
   e. to cause membranes to fluoresce
   answer: b

5. Sphingomyelin would not have which of the following functional groups?
   a. amide
   b. disubstituted phosphate
   c. hydroxyl
   d. sulfate
   answer: d

6. Which of the following lipids has the greatest molecular weight?
   a. diphosphatidyl glycerol
   b. phosphatidyl serine
   c. sphingomyelin
   answer: a

7. Which of the following is not a true statement about phospholipides?
    a. all are charged
    b. double bonds in the fatty acyl groups are usually in the trans configuration
    c. the molecular identity of the C-1 and C-2 fatty acyl groups are usually different
    d. they are major constituents of membranes in bacteria and plants
    answer: b

8. Which term describes the dual nature of phospholipids, possessing structures conferring solubility in water and other structures hydrophobic in nature?
    a. amphipathic
    b. schizophrenic
    c. schizothymic
    d. anpleurotic
    e. bipolar
    answer: a

9. Which of the following is a false statement about micelles?
    a. they are in equilibrium with the monomeric form of the detergent in solution
    b. they are free to rotate in solution
    c. they are formed because the polar head groups cause the formation of netlike structures of highly-ordered water molecules
    d. they contain almost no water
    answer: c

10. Which of the following is a false statement about liposomes?
    a. they may contain an aqueous center
    b. the molecules forming liposomes usually have polar head groups
    c. their form is usually spherical
    d. their surface charge can depend on the pH of the suspending medium
    e. none of the above
    answer: e

11. Membranes appear trilaminar in the electron microscope because
    a. membranes fluorescence gives off three photons for each electron
    b. phospholipid head groups are more electron dense than the hydrophobic tails
    c. membranes are composed of three layers: lipid, protein and lipid
    d. each membrane is coated with a layer of protein on each side usually forming a dense β-sheet
    answer: b

12. Which of the following is true about peripheral membrane proteins?
    a. they are covalently attached to phospholipids such as phosphatidyl serine by amide bonds
    b. they can be separated from other membrane components by using high concentrations of synthetic detergents
    c. they are not denatured by boiling
    d. they can be removed from membranes with high concentrations of NaCl
    e. they do not contain hydrophobic amino acid residues
    answer: **d**

13. Analysis of an integral membrane protein's primary structure reveals that it has a hydropathy plot with two strongly positive peaks spanning 22-27 residues each. What could you deduce from this information?
    a. there are two membrane-spanning chains which must be anti-parallel to each other
    b. there are likely three membrane spanning helixes
    c. each positive peak on the hydropathy plot spans the membrane twice
    d. the N-terminus and the C-terminus are on opposite sides of the membrane
    answer: a

14. Which is true about the membrane/aqueous interface?
    a. phospholipids are lined up with the hydrocarbon chains extended out into the aqueous environment
    b. the interface involves only proteins and not lipids
    c. the membrane surface charges change with changing pH
    d. membrane formation is endothermic, so ice crystals are present
    answer: c

15. Bacteriorhodopsin is a model membrane protein with seven membrane spanning sequences. Which of the following is true?
    a. the N-terminus and C-terminus are on the same side of the membrane
    b. the protein must have a molecular weight greater than 100,000
    c. it has a large proportion of its structure as β-pleated sheet
    d. it is unlikely to contain any histidine residues
    e. it uses the energy from photons to pump protons across the membrane
    answer: e

16. The carbohydrate residues on membrane glycoproteins are shielded from the hydrophobic interior of the membrane by what adaptation?
    a. they are phosphorylated
    b. they are buried at the very center of the membrane in the space between the hydrophobic tails of the phospholipids
    c. they are buried inside the tertiary structure of the protein
    d. they stick out into the aqueous phase and are not shielded
    answer: d

17. Higher plant chloroplast thylakoid membranes are different from other membranes in that the major lipid is not a phospholipid, but a galactolipid. Rather than a phosphate esterified to the C-3 of a diacylglycerol, galactolipids have a mono or digalactoside linkage to the C-3 position. What effect might this difference in the lipid composition produce?
   a. the major carbon source for glycolysis in plants will be galactose, not glucose
   b. thylakoid membranes would be tetralaminar
   c. O_2 and CO_2 will be more accessible to the hemoglobin inside the chloroplast
   d. the surface charge on the thylakoid membrane would be more influenced by the protein composition than is the case for most microbial or animal membranes
   answer: d

18. Diphosphatidylglycerol is actually two phosphatidic acid molecules, each esterified to opposite ends of a glycerol molecule. Thus it produces four fatty acids, two phosphates and three glycerols upon hydrolysis. Which of the following is true about its localization in the membrane?
   a. each diphosphatidylglycerol molecule spans the entire membrane and interacts with both layers of the lipid bilayer
   b. each diphosphatidylglycerol molecule will insert one phosphatidyl group in a membrane lipid layer and the other phosphatidyl group will stick out in the membrane/air interface
   c. each diphosphatidylglycerol molecule will reside on only one side of the bilayer
   d. diphosphatidylglycerol molecule will probably diffuse faster in the membrane than phosphatidic acid molecule with a similar fatty acyl composition
   answer: c

19. Differential scanning calorimetry is used as a physical probe to study membranes. This technique will provide information about all but which of the following?
   a. the average fatty acyl chain length of the lipids
   b. the temperature at which the microbe producing the membrane was cultured
   c. the relative $Na^{\varnothing}$ and $K^{\varnothing}$ concentrations in the cell
   d. the relative proportion of cholesterol in the membrane
   e. the presence of an unusually high number of cis double bonds in the lipids
   answer: c

20. The most common type of phospholipids do not contain which of the following?
   a. an alcohol
   b. phosphate
   c. glycerol
   d. a purine
   e. fatty acids
   answer: d

21. DNA polymerase is used as an organelle-specific enzyme marker for
    a. lysosomes
    b. nuclei
    c. golgi apparatus
    d. endoplasmic reticular vesicles
    e. none of the above
    answer: b

22. Glycosyl transferase is used as an organelle-specific enzyme marker for
    a. lysosomes
    b. nuclei
    c. golgi apparatus
    d. endoplasmic reticular vesicles
    e. none of the above
    answer: c

23. The inner and outer mitochondrial membranes can be distinguished by the presence of
    a. monoamine oxidase in the inner and cytochrome c in the outer
    b. monoamine oxidase in the outer and cytochrome c in the inner
    c. acid phosphatase in the inner and cytochrome c in the outer
    d. acid phosphatase in the outer and cytochrome c in the inner
    answer: b

24. Membrane structures are commonly separated by their
    a. size
    b. viscosity
    c. melting transition
    d. density
    e. charge
    answer: d

25. A polyunsaturated fatty acid
    a. stearic acid
    b. oleic acid
    c. arachidonic acid
    d. palmitic acid
    e. myristic acid
    answer: c

26. The fluidity of a membrane
    a. is determined by differential scanning calorimetry
    b. depends on the average chain length of its fatty acids
    c. is determined by sucrose density centrifugation
    d. a and b
    e. b and c
    answer: b

27. Lipid components of membranes readily move about in one side of a lipid bilayer.
    a. true
    b. false
    answer: a

28. Lipid components of membranes do not readily move from one side of a bilayer to the other.
    a. true
    b. false
    answer: a

29. If lipid A has a lower Tm than lipid B then at the same temperature lipid B has a greater fluidity than lipid A.
    a. true
    b. false
    answer: b

30. Comment on the fatty acyl chain length and degree of unsaturation on the Tm of the corresponding lipid.
    Answer: As the chain length increases and the number of unsaturations decreases the Tm increases.

31. What, if any, impact would you expect to find on the fatty acid content of the lipids from *E. coli* grown at 25 °C versus *E. coli* grown at 40 °C?

    Answer: Those grown at 25 °C would be expected to have fatty acids with shorter chain lengths and/or more unsaturations.

Chapter 20 Mechanisms of Membrane Transport

1. If 10 mL of a solution containing 1 mM D-glucose and 1mM L-glucose in dialysis tubing is placed in a beaker with 100 mL of 1 mM D-glucose, which of the following will occur?
   a. L-glucose concentration inside the dialysis tubing will remain relatively constant
   b. D-glucose concentrations inside the dialysis tubing will increase
   c. the concentration of neither carbohydrate will change until ATP is added to the solution in the beaker
   d. the concentration of neither carbohydrate will change until ATP is added to the solution inside the dialysis tubing
   e. D-glucose will move across the membrane in both directions
   answer: e

2. The outer membrane of Gram-negative bacteria, mitochondria and chloroplasts are naturally leaky. Of the following molecules, which would be the largest able to pass through these membranes?
   a. glucose
   b. glycine
   c. hemoglobin
   d. lactose
   e. water
   answer: d

3. When cells of *E. coli* are placed in a solution with both D-glucose and lactose, the cells will preferentially transport and utilize the D-glucose until its concentration is depleted and then will start to import and catabolize the lactose. This phenomenon is called "diauxie." What is a likely reason for this phenomenon?
   a. D-glucose is a competitive inhibitor of the lactose transport system
   b. the lactose transport system is not produced until the glucose is consumed
   c. carbohydrates cannot be transported until glucose catabolism produces ATP in the cells
   d. the cell membrane contains an antiport system which pumps lactose out while transporting D-glucose into the cell
   answer: b

4. Which of the following is a true statement about facilitated diffusion?
   a. such systems are generally capable of transport in only one direction
   b. such systems are composed of peripheral membrane proteins
   c. such systems will still transport solute molecules across the membrane when C_{in} = C_{out}
   d. increasing the concentration of the transported solute on one side of the membrane proportionally increases the rate of transport to the other side of the membrane
   answer: c

5. Which of the following is not an important component of membrane transport in cells?
    a. $\Delta\Psi$ (membrane surface potential)
    b. ΔC (solute concentration)
    c. $\Delta G^{o'}$ (standard free energy)
    d. ΔH^+ (proton concentration)
    e. ΔP (pressure)
    answer: e

6. Which is likely true about the affinity of the *E. coli* lactose permease for lactose?
    a. >1 M
    b. higher on the inside of the membrane than on the outside
    c. lower on the outside of the membrane than on the inside
    d. different on the inside and the outside of the membrane
    answer: d

7. All but which of the following could be used to drive transport by an electrogenic pump?
    a. Q cycle
    b. sodium potassium ATPase
    c. photons
    d. sodium dodecylsulfate
    answer: d

8. Which of the following is not commonly used to study transport processes?
    a. membrane vesicles
    b. substrate analogs
    c. ampholytes
    d. mutants
    e. isotopes
    answer: c

9. At one point in the mechanism of the sodium-potassium pump a phosphoryl group from ATP is transferred to a protein aspartyl group. The product formed is
    a. an ester
    b. an asparaginyl group
    c. an anhydride
    d. AMP
    e. PPi
    answer: c

10. Which of the following would you expect to be most likely to be transported by lactose permease?
    a. maltose
    b. sucrose
    c. galactose
    d. β-phenolgalactoside
    e. α-phenolgalactoside
    answer: d

11. Lactose permease has
    a. a β-barrel shape
    b. a significant antiparallel β-sheet contribution
    c. a helix-turn-helix motif
    d. a helix-loop helix motif
    e. 12-transmembrane helices
    answer: e

12. Which of the following requires energy?
    a. active transport
    b. diffusion
    c. facilitated diffusion
    d. transport of charged molecules
    e. transport of zwitterions
    answer: a

13. Which of the following is not true for the *Rhodobacter* porin?
    a. it is a trimer
    b. each molecule forms a tube that passes through the membrane
    c. is a β-barrel with 16 antiparallel β-strands
    d. allows the passage of molecules up to molecular weight of 1500
    e. are in the outer membrane
    answer: d

14. *o*-Nitrophenyl-β-galactoside is used to
    a. study the assembly of synthetic vesicles
    b. study lactose permease
    c. produce hydropathy profiles for membrane proteins
    d. produce osmotic lysis
    e. study the parallel bundling of helices
    answer: b

15. Bacterial inner membranes are more permeable than the outer membranes.
    a. true
    b. false
    answer: b

125

16. *E. coli* growing on glucose has lactose transporters in its plasma membrane.
    a. true
    b. false
    answer: b

17. An antiport transports two different solutes to opposite sides of a membrane.
    a. true
    b. false
    answer: a

18. For the eukaryotic sodium-potassium pump, each ATP hydrolyzed moves 2 sodium ions out and three potassium ions into the cell.
    a. true
    b. false
    answer: b

19. The receptor for insulin is a tyrosine kinase.
    a. true
    b. false
    answer: a

20. The toxin from cholera catalyzed the transfer of an AMP moiety from NAD^+ to a specific arginyl group on a subunit of the β-adrenergic receptor.
    a. true
    b. false
    answer: b

21. Which of the following is not true about gap junctions?
    a. they are composed of rod shaped proteins
    b. they are hexagonal arrays
    c. they consist of a trimer, with each protein being a β-barrel with 16 antiparallel β-strands
    d. the channels from one cell line up with those of adjacent cells
    e. they allow small molecules to be transmitted from cell to cell
    answer: c

22. Which of the following is true for the *Rhodobacter* porin?
    a. it is a trimer protein that forms one transmembrane tube
    b. it is a trimer protein that forms three transmembrane tubes
    c. it is a monomer protein that forms a transmembrane tube
    d. it does not have a transmembrane tube
    answer: b

23. The insulin receptor consists of (X) membrane spanning β-chain(s) linked by disulfide bonds to (Y) chain(s) on the outer surface of the plasma membrane.

	(X)	(Y)
a.	1	1
b.	1	2
c.	2	2
d.	2	1
e.	3	1

answer: c

24. The insulin receptor kinase is located in the X chains and promotes the transfer of a phosphoryl group from an ATP to a Y on the Z chain.

	X	Y	Z
a.	α	tyrosine	β
b.	β	tyrosine	β
c.	β	serine	α
d.	β	serine	β
e.	β	tyrosine	α

answer: b

25. Describe how the substrate analog o-nitrophenyl-β-galactoside functions as a tool to study transport rates etc.

answer: The colorless o-nitrophenyl-β-galactoside is transported by lactose permease to the interior of the cell where it is hydrolyzed by β-galactosidase and the yellow o-nitrophenol is formed and quantified spectrophotometrically.

Chapter 21 Metabolism of Fatty Acids

1. The combination of a triacylglycerol with pancreatic lipase will produce which of the following?
    a. monoacylglycerol + diacylglycerol + fatty acids
    b. glycerol + phosphate + fatty acids
    c. glycerol + fatty acids
    d. glyceraldehyde + fatty acids
    e. diacylglycerol + fatty acids
    answer: a

2. Spherical particles used to transport lipids in the lymph and bloodstream are called
    a. micelles
    b. liposomes
    c. chylomicrons
    d. plastoglobuli
    e. lipocelles
    answer: c

3. The major energy demands of heart tissue and brain tissue are provided by which carbon sources?
    a. glucose
    b. ketone bodies
    c. fatty acids and glucose
    d. fatty acids
    e. glucose and glutamate
    answer: c

4. Dietary fats (triglycerides) are imported into the body by which tissues?
    a. adipose tissue
    b. intestine
    c. liver
    d. pancreas
    e. skin
    answer: b

5. Both wood (cellulose) and wax (the ester of a fatty acid and a fatty alcohol) will burn (react with oxygen to produce water and carbon dioxide). Which of the following is true about these two processes?

    a. burning the wax will produce light energy, but the burning of wood will not produce light

    b. based on an equal amount of carbon dioxide produced, the wood will produce more energy

    c. the carbons in the wood are more oxidized than those in the wax

    d. wood will only burn in the presence of oxygen, while wax can burn under anaerobic conditions

    answer: c

6. The reaction catalyzed by this enzyme produces enoyl-CoA

    a. acyl-CoA dehydrogenase

    b. enoyl-CoA hydrase

    c. HMG-CoA lyase

    d. HMG-CoA synthase

    e. β-L-hydroxyacyl-CoA dehydrogenase

    answer: a

7. The reaction catalyzed by this enzyme produces NADH

    a. acyl-CoA dehydrogenase

    b. enoyl-CoA hydrase

    c. HMG-CoA lyase

    d. HMG-CoA synthase

    e. β- L-hydroxyacyl-CoA dehydrogenase

    answer: e

8. The reaction catalyzed by this enzyme produces one of the ketone bodies

    a. acyl-CoA dehydrogenase

    b. enoyl-CoA hydrase

    c. HMG-CoA lyase

    d. HMG-CoA synthase

    e. β- L-hydroxyacyl-CoA dehydrogenase

    answer: c

9. The reaction catalyzed by this enzyme produces $FADH_2$

    a. acyl-CoA dehydrogenase

    b. enoyl-CoA hydrase

    c. HMG-CoA lyase

    d. HMG-CoA synthase

    e. β- L-hydroxyacyl-CoA dehydrogenase

    answer: a

10. The reaction catalyzed by this enzyme produces β- L-hydroxyacyl-CoA
    dehydrogenase
    a. acyl-CoA dehydrogenase
    b. enoyl-CoA hydrase
    c. HMG-CoA lyase
    d. HMG-CoA synthase
    e. β- L-hydroxyacyl-CoA dehydrogenase
    answer: b

11. The reaction catalyzed by this enzyme produces β-ketoacyl-CoA
    a. acyl-CoA dehydrogenase
    b. enoyl-CoA hydrase
    c. HMG-CoA lyase
    d. HMG-CoA synthase
    e. β- L-hydroxyacyl-CoA dehydrogenase
    answer: e

12. This molecule is the product of an enzyme which uses biotin as a coenzyme
    a. palmitoyl-CoA
    b. acetyl-CoA
    c. acetoacetyl-CoA
    d. malonyl-CoA
    e. linoleoyl-CoA
    answer: d

13. Which of the following molecules has the greatest number of unsaturated bonds?
    a. palmitoyl-CoA
    b. malonyl-CoA
    c. linoleoyl-CoA
    d. palmitoleoyl-CoA
    e. acetoacetyl-CoA
    answer: c

14. This molecule will produce the most energy from β-oxidation
    a. palmitoyl-CoA
    b. malonyl-CoA
    c. acetyl-CoA
    d. palmitoleoyl-CoA
    e. acetoacetyl-CoA
    answer: a

15. This molecule is a feed-back inhibitor of acetyl-CoA carboxylase
    a. palmitoyl-CoA
    b. acetyl-CoA
    c. linoleoyl-CoA
    d. palmitoleoyl-CoA
    e. acetoacetyl-CoA
    answer: a

16. This molecule is used to activate free fatty acids prior to β-oxidation
    a. coenzyme A
    b. acyl carrier protein
    c. biotin
    d. carnitine
    e. phosphopantotheine
    answer: a

17. This enzyme is a product of the thiolase enzyme-catalyzed reaction
    a. coenzyme A
    b. acyl carrier protein
    c. biotin
    d. carnitine
    e. phosphopantotheine
    answer: a

18. This molecule is a product of the thioesterase enzyme-catalyzed reaction
    a. coenzyme A
    b. acyl carrier protein
    c. biotin
    d. carnitine
    e. phosphopantotheine
    answer: b

19. This molecule is used to transport fatty acids across the mitochondrial inner membrane
    a. coenzyme A
    b. acyl carrier protein
    c. biotin
    d. carnitine
    e. phosphopantotheine
    answer: d

20. This molecule is derived from a vitamin and used to produce the two substrates for the acetyl-CoA-ACP transacylase enzyme
    a. coenzyme A
    b. acyl carrier protein
    c. biotin
    d. carnitine
    e. phosphopantotheine
    answer: e

21. The complete β-oxidation of palmitoyl-CoA will require how many cycles through the pathway?
    a. seven
    b. eight
    c. nine
    d. ten
    e. sixteen
    answer: a

22. The β-oxidation of oleoyl-CoA ($18:1\Delta^9$) requires the additional activity of which type of enzyme relative to the β-oxidation of palmitoyl-CoA?
    a. dehydrogenase
    b. isomerase
    c. ligase
    d. lyase
    e. hydrolase
    answer: b

23. Which of the following statements about ketone bodies is incorrect?
    a. ketone bodies are produced primarily in the liver
    b. ketone bodies can be used by brain tissue as an energy source
    c. the $\Delta^{o'}$ for acetoacetyl-CoA thiolase, the first enzyme in the pathway for ketone body synthesis, is positive
    d. the noticeable presence of ketone bodies in the breath of diabetics indicates that too much glucose is being catabolized
    answer: d

24. Which of the following statements about acetyl-CoA carboxylase is incorrect?
    a. it catalyzes the first step in the synthesis of fatty acids
    b. it utilizes biotin as a coenzyme
    c. the product of the reaction contains three carbons
    d. it is activated by citrate
    e. the activated form of the enzyme is a large polymer
    answer: c

25. If the numbering of fatty acids is such that the carboxyl carbon is number 1, which carbon in palmitic acid is oxidized during the first cycle of β-oxidation?
    a. 1
    b. 2
    c. 3
    d. 4
    e. 15
    answer: c

26. The animal fatty acid synthase is a single multifunctional polypeptide able to carry out the seven reactions of the pathway. Which of the following is an advantage of this arrangement over the production of seven different enzymes?
    a. the Km for the first enzyme in the pathway will be lower
    b. the ACP can be covalently attached
    c. it is easier to run the reactions in either the forward (biosynthesis) or reverse (catabolism) direction, enhancing cellular regulation
    d. dilution of intermediates in the pathway is minimized
    answer: d

27. In which organism would you expect to find the anaerobic pathway for production of monosaturated fatty acids?
    a. bacteria
    b. fish
    c. mammals
    d. plants
    e. birds
    answer: d

28. Cats lack the ability to produce arachidonic acid from linoleic acid. What is the result of this deficiency?
    a. cats must eat plants to stay healthy
    b. cats must consume milk to provide a source of lactose
    c. arachidonic acid is an essential fatty acid for cats
    d. cats do not produce wax in their ears
    answer: c

29. Triglyciderides in biochemistry teachers are stored primarily in which tissue?
    a. adipose
    b. brain
    c. liver
    d. muscle
    e. heart
    answer: a

30. Fatty acid synthesis
    a. occurs in the mitochondrion in animal cells
    b. utilizes FAD and NAD^+ as electron donors
    c. is necessary for triacylglycerol biosynthesis
    d. is regulated independently of gluconeogenesis
    answer: c

31. The activity of triacylglycerol lipase by glucagon involves the consumption of ATP. Which two enzymes involved in this regulation utilize ATP as a substrate?
    a. hexokinase and pyruvate kinase
    b. adenylate cyclase and protein kinase
    c. citrate synthase and ATPase
    d. F_o and F_1
    e. acyl-CoA ligase and lipokinase
    answer: b

32. The immediate precursor of the fatty acyl groups in triglycerides are
    a. fatty acid phosphate mixed anhydrides
    b. fatty acids
    c. fatty acyl-CoA's
    d. fatty amides
    answer: c

33. In the conversion of propionyl-CoA to succinyl-CoA, why didn't nature convert propionyl-CoA directly into succinyl-CoA?
    a. the carboxylase could have evolved from acetyl-CoA carboxylase
    b. the biotin mechanism requires a resonance stabilized carbanion
    c. methylmalonyl-CoA is a metabolite in an amino acid catabolic pathway in anaerobes and could have contributed evolutionarily to the pathway
    d. a, b, and c
    e. none of the above
    answer: d

34. The conversion of acetoacetate to acetone
    a. might be non-enzymatic
    b. is facilitated by a resonance stabilized carbanion
    c. changes one ketone body into another
    d. occurs in the mitochondria
    e. all of the above
    answer: e

35. Which is true for fatty acid biosynthesis
   a. involves NADH, CO_2, acetyl-CoA, carnitine
   b. involves NADPH, CO_2, malonyl-CoA, ACP
   c. involves NADH, CO_2, acetyl-CoA, malonyl-CoA
   d. involves NADH, $FADH_2$, CO2, acetyl-CoA, carnitine
   e. involves NADPH, $FADH_2$, ACP, acetyl-CoA
   answer: b

36. Produce a formula that predicts the number of ATP's produced per n where n is the number of carbons in an even numbered mitochondrial fatty acyl-CoA. Use β-oxidation, the citric acid cycle, and electron transport.

   answer: $4 (n/2 - 1) + 10(n/2) = 7n - 4$

Chapter 22   Biosynthesis of Membrane Lipids

1. Regardless of the organism involved, biosynthetic reactions for phospholipids generally occur
    a. in the Golgi apparatus
    b. mainly on the surface of membranes
    c. in the cytosolic fluid
    d. on the endoplasmic reticulum
    answer: b

2. Because of their physical properties, many of the enzymes that catalyze phopholipid biosynthesis may be described as
    a. hydrophobic
    b. hydrophilic
    c. amphipathic
    d. electronically neutral
    answer: c

3. The nitrogenous base(s) in the high-energy nucleotide derivatives of the intermediates in the biosynthesis of phospholipids is
    a. adenine and cytosine
    b. adenine
    c. guanine and uracil
    d. cytosine
    e. uracil
    answer: d

4. In *E. coli* the last common intermediate for the synthesis of phosphatidylethanolamine and phosphatidylglygerol is
    a. diacylglycerol
    b. CDP-diacylglycerol
    c.   phosphatidylserine
    d. phosphatidic acid
    answer: b

5. In phospholipid synthesis by *E. coli*, glycerol-3-phosphate acyltransferase, shows higher enzymatic activities when the carbon chain of the acyl derivative that serves as a substrate has
    a. 17 carbons or more
    b. less than 15 carbons
    c. no double bonds
    d. at least one double bond
    answer: c

6. In lung tissue, the fatty acyl substituents of phosphatidylcholine may be altered (changed from one fatty acyl group to another) by
    a. deacylation of the acyl moiety followed by reacylation with the desired acyl moiety
    b. conversion of the fatty acyl group while it is attached to the remainder of the phosphatidylcholine molecule
    c. direct exchange of the fatty acyl groups at the SN-1 and SN-2 positions
    d. re-synthesis of the desired diacylglycerol followed by condensation with CDP-choline
    answer: a

7. In the phospholipids of *E. coli*, most of the acyl residues at the SN-2 position are unsaturated because during their synthesis
    a. 1-acylglycerol-3-phosphate acyltransferase shows a preference for acyl derivatives that are unsaturated
    b. glycerol-3-phosphate acyltranferase shows a preference for acyl derivatives that are saturated
    c. only unsaturated residues are transferred to the glycerol moieties, then the acyl residues occupying the SN-1 positions are reduced with NADH
    d. the phosphatidic acid intermediates that have saturated acyl residues at their NS2 positions are acted upon by phosphatidic acid dehydrogenase
    answer: a

8. The substrates for the final step in the biosynthesis of diphosphatidylglycerol by *E. coli* are
    a. CDP-diacylglycerol and phosphatidylglycerol
    b. phosphatidylglycerol phosphate and phosphatidylglycerol
    c. two molecules of CDP-diacylglycerol
    d. two molecules of phosphatidylglycerol
    answer: d

9. Diphosphatidylglycerol is synthesized in the mitochondria of eukaryotes by the condensation of
    a. two molecules of phosphatidylglycerol
    b. two molecules of CDP-diacylglycerol and elimination of two molecules of CMP
    c. CDP-diacylglycerol with phosphatidylglycerol and elimination of CMP
    d. phosphatidylglycerol phosphate with phosphatidylglycerol and elimination of phosphate
    answer: c

10. In the biosynthesis of phosphatidic acid by eukaryotes, 1-acylglycerol-3-phosphate may be formed by
   a. the deacylation of phosphatidic acid at the SN-1 position
   b. the acylation of glycerol-3-phosphate
   c. the reduction of 1-acyldihydroxyacetone phosphate with NADPH
   d. b and c
   answer: d

11. Phosphatidylinositol
   a. only functions as an intermediate in the biosynthesis of of phosphatidylinositol-4 phosphate in eukaryotes
   b. does not occur in the membranes of eukaryotes
   c. constitutes about 5 % of the phospholipids in prokaryotes
   d. may be converted to two phosphatidylinositol derivatives that yield cellular second messengers in animal cells
   answer: d

12. When activated by a hormone binding to a cell's surface, phosphatidylinositol-4,5-bisphosphate in the cell's membrane may be hydrolyzed by a phospholipase to yield two second messengers, namely
   a. inositol-4,5-bisphosphate and phosphatidic acid
   b. inositol-1,4,5-triphosphate and diacylglycerol
   c. inositol-1,4,5-triphosphate and phosphatidic acid
   d. triacylglycerol and inositol-4,5-bisphosphate
   answer: b

13. In eukaryotes, phosphatidylserine is synthesized by the reaction of
   a. serine with CDP-diacylglycerol
   b. CDP-serine with diacylglycerol
   c. serine with phosphatidylethanolamine
   d. CDP-serine with phosphatidic acid
   e. phosphatidylethanolamine and CO_2
   answer: c

14. Which of the following is not a valid statement concerning the biosynthesis of phosphatidylethanolamine?
   a. In eukaryotes, phosphatidylserine may react with ethanolamine to yield phosphatidylethanolamine.
   b. In both prokaryotes and eukaryotes, phosphatidylserine may be decarboxylated to yield phosphatidylethanolamine and carbon dioxide.
   c. In prokaryotes, phosphatidylserine is the only immediate precursor of phosphatidylethanolamine.
   d. In both prokaryotes and eukaryotes, CDP-ethanolamine reacts with diacylglycerol to form phosphatidylethanolamine and CMP.
   answer: d

15. The enzyme CTP-phosphocholine cytidyltransferase
    a. is located in the plasma membrane in *E. coli*
    b. may exist in an inactive form in the cytosol of eukaryotic cells
    c. is inactivated by high levels of diacylglycerol in the endoplasmic reticulum
    d. remains bound to the endoplasmic reticulum, but is activated by dephosphorylation of the enzyme
    answer: b

16. The synthesis of phosphatidylcholine in eukaryotes is
    a. stimulated by a decrease in phosphatidylcholine concentration and an increase in the concentration of diacylglycerol in the endoplasmic reticulum membranes
    b. inhibited by dephosphorylation of CDP-phosphocholinecytidyltransferase
    c. only controlled by the level of CTP-phosphocholine cytidyltransferase
    d. is regulated by the allosteric enzyme phosphatidate cytidyltranferase, for which choline is a positive effector
    answer: a

17. Which enzymatic activity yields choline and phosphatidic acid when phosphatidylcholine is the substrate?
    a. phospholipase A_1
    b. phospholipase A_2
    c. phospholipase C
    d. phospholipase D
    e. lysophospholipases
    answer: d

18. Which enzymatic activity hydrolyzes the ester linkage at the SN-1 position of phospholipids?
    a. phospholipase A_1
    b. phospholipase A_2
    c. phospholipase C
    d. phospholipase D
    e. lysophospholipases
    answer: a

19. Which enzymatic activity hydrolyzes a phospholipid that lacks an acyl residue at either the SN-1 or SN-2 position?
    a. phospholipase A_1
    b. phospholipase A_2
    c. phospholipase C
    d. phospholipase D
    e. lysophospholipases
    answer: e

20. Which enzymatic activity yields phosphoserine and diacylglycerol when phosphatidylserine is the substrate?
    a. phospholipase A_1
    b. phospholipase A_2
    c. phospholipase C
    d. phospholipase D
    e. lysophospholipases
    answer: c

21. Which enzymatic activity removes the acyl residue from the SN-2 position of a phospholipid?
    a. phospholipase A_1
    b. phospholipase A_2
    c. phospholipase C
    d. phospholipase D
    e. lysophospholipases
    answer: b

22. The major phospholipase found in most snake venoms hydrolyzes
    a. the linkage between the glycerol and phosphate moieties
    b. the ester linkage at the SN-2 position
    c. both of the ester linkages
    d. the linkage between the phosphate and the head group moiety
    e. all of the linkages to the glycerol moiety
    answer: b

23. The initial step in the synthesis of sphinganine is the reaction of pamitoyl-CoA with
    a. choline
    b. serine
    c. ethanolamine
    d. lactate
    e. glycerol
    answer: b

24. The final step in the synthesis of sphingomyelin involves the transfer of a phosphocholine group to ceramide. The donor of the phosphocholine group is
    a. phosphatidylcholine
    b. CTP-choline
    c. phosphorylcholine
    d. choline and ATP
    e. none of the above
    answer: a

25. Glycosphingolipids found in the plasma membrane are oriented
    a. symmetrically in the bilayer with the glycosyl residues evenly distributed on the surface of both sides
    b. asymmetrically in the bilayer with sphingolipids residing in the layer next to the interior of the cell
    c. asymmetrically in the bilayer with the glycosyl residues facing the outside of the cell
    d. symmetrically in the bilayer with the glycosyl residues located in the interior of the bilayer
    answer: c

26. Some glycosphingolipids have a structural role but others appear to be involved as
    a. second messengers
    b. energy storage molecules
    c. hormone receptors
    d. cell surface recognition factors
    answer: d

27. The catabolic enzymes acting on sphingolipids are located in the
    a. plasma membrane
    b. lysosomes
    c. mitochondria
    d. endoplasmic reticulum
    e. nucleus
    answer: b

28. The fatty acid precursor for the synthesis of eicosanoid hormones is stored as
    a. free fatty acid in tissues
    b. an ester in the SN-1 and Sn-2 positions of triacylglycerols
    c. an ester in the SN-2 position of phospholipids
    d. none of the above
    answer: c

29. Prostaglandins affect target cells by
    a. binding to specific cell surface receptors that activate adenylate cyclase within the cell
    b. entering a cell and inhibiting a specific protein kinase that regulates glucose metabolism
    c. binding to hormone-like cell surface receptors that activate specific phospholipases
    d. entering cells and serving as a positive allosteric effector for the enzyme
    answer: a

30. During the synthesis of 3-ketosphinganine
   a. the carboxyl carbon of a fatty acid is lost as carbon dioxide
   b. the carboxyl carbon of serine is lost as carbon dioxide
   c. an ester bond is formed
   d. a succinoyl-CoA combines with a glycine
   e. the carboxyl carbon of alanine is lost as carbon dioxide
   answer: b

31. Choline is an essential nutrient for humans.
   a. true
   b. false
   answer: a

32. The biosynthesis of alkyl ether lipids involves the reduction of ester linkages.
   a. true
   b. false
   answer: b

33. Like unsaturated fatty acids, ceramides have the cis geometry.
   a. true
   b. false
   answer: b

34. The S in SAM stands for saturated.
   a. true
   b. false
   answer: b

Chapter 23 Metabolism of Cholesterol

1. The carbon skeleton of cholesterol is derived from
   a. pyruvate
   b. butyrate
   c. isovalerate
   d. acetate
   e. malonate
   answer: d

2. The substrates for the reaction catalyzed by HMG-CoA synthase are
   a. two molecules of acetyl-CoA
   b. malonyl-CoA and β-ketobutyryl-CoA
   c. acetyl-CoA and β-ketobutyryl-CoA
   d. acetyl-CoA and malonate
   e. β-hydroxy-β-methylglutaryl-CoA and malonate
   answer: c

3. In animals, HMG-CoA is synthesized
   a. in the cytosol and mitochondrial matrix
   b. on the endoplasmic reticulum and plasma membrane
   c. in the Golgi apparatus primarily
   d. on the endoplasmic reticulum and outer mitochondrial membrane
   e. on the plasma membrane
   answer: a

4. HMG-CoA is converted to ketone bodies when it is synthesized
   a. on the endoplasmic reticulum
   b. in the mitochondria
   c. in the Golgi apparatus
   d. in the cytosol
   e. on the plasma membrane
   answer: b

5. Several reactions are involved in the control of cholesterol biosynthesis but the primary regulation site appears to be
   a. thiolase in the mitochondria
   b. mebalonate kinase on the endoplasmic reticulum
   c. HMG-CoA reductase on the endoplasmic reticulum
   d. HMG-CoA synthase in the cytosol
   answer: c

143

6. Which of the following statements about cholesterol biosynthesis control mechanisms associated with HMG-CoA reductase is false?
    a. the amount of HMG-CoA reductase mRNA is regulated by the concentration of cholesterol
    b. the rate of degradation of HMG-CoA reductase is modulated by the supply of cholesterol
    c. HMG-CoA reductase in inactivated by phosphorylation and activated by dephosphorylation
    d. cholesterol serves as a negative allosteric effector for HMG-Co reductase
    answer: d

7. Lovastatin acid is a drug that is effective in treating hypercholesterolemia because it
    a. is a competitive inhibitor of HMG-CoA reductase
    b. stimulates the production of high-density lipoproteins
    c. stimulates the conversion of cholesterol to bile acids
    d. is a noncompetitive inhibitor of isopentenyl pyrophosphate isomerase
    answer: a

8. An intermediate in the cyclization of squalene to form lanosterol is
    a. squalene-2,3-oxide
    b. squalene pyrophosphate
    c. demosterol
    d. squalene phosphate
    e. a structure having 3 six-member and 1 five-member rings fused together and containing 28 carbon atoms
    answer: a

9. Which of the following has the highest percentage of triacylglycerol
    a. chylomicron
    b. very low-density lipoprotein
    c. intermediate-density lipoprotein
    d. low-density lipoprotein
    e. high density lipoprotein
    answer: a

10. The enzyme lecithin cholesterolacyltransferase (LACT) is associated with this lipoprotein in the plasma.
    a. chylomicron
    b. very low-density lipoprotein
    c. intermediate-density lipoprotein
    d. low-density lipoprotein
    e. high density lipoprotein
    answer: e

11. The level of this lipoprotein is elevated in patients suffering from familial hypercholesterolemia.
    a. chylomicron
    b. very low-density lipoprotein
    c. intermediate-density lipoprotein
    d. low-density lipoprotein
    e. high density lipoprotein
    answer: d

12. This lipoprotein is synthesized by intestinal cells.
    a. chylomicron
    b. very low-density lipoprotein
    c. intermediate-density lipoprotein
    d. low-density lipoprotein
    e. high density lipoprotein
    answer: a

13. This lipoprotein has the highest percentage of protein.
    a. chylomicron
    b. very low-density lipoprotein
    c. intermediate-density lipoprotein
    d. low-density lipoprotein
    e. high density lipoprotein
    answer: e

14. This lipoprotein has the smallest diameter of the lipoprotein types.
    a. chylomicron
    b. very low-density lipoprotein
    c. intermediate-density lipoprotein
    d. low-density lipoprotein
    e. high density lipoprotein
    answer: e

15. In the human lipoproteins, the apoproteins are located
    a. in the core of a micellar-like structure
    b. in the phospholipid-cholesterol monolayer
    c. on the outer surface of the spherical lipid bilayer
    d. with the triacylglycerol that is present within the structure
    e. none of the above
    answer: b

16. The tissues primarily responsible for the removal of low-density lipoproteins from plasma are
     a. gonads, adrenal glands , and kidney
     b. kidney, gallbladder, and intestinal tissue
     c. liver, kidney, and heart
     d. liver, adrenal glands, and adipose tissue
     e. heart, gonads, and gallbladder
     answer: d

17. The uptake of low-density lipoprotein (LDL) by cells involve several processes. Which of the following descriptions of those processes is not correct?
     a. The protein called B100 binds to one of the cell's LDL receptors which are located in regions of the plasma membrane called pits.
     b. Inside the cell, coated vesicles containing LDL fuse with lysosomes.
     c. When degraded, the LDL particles are dissembled into lipids and apoproteins, which are recycled without further degradation.
     d. LDL particles enter the cell along with their receptors by endocytosis.
     answer: c

18. Patients with the homozygous form of the disease called familial hypercholesterolemia exhibit
     a. atherosclerosis, in spite of elevated levels of high-density lipoprotein levels
     b. a mortality rate that is consistent with a lifetime of 40 years
     c. elevated plasma cholesterol levels but normal low-density lipoprotein levels
     d. xanthomas, which are cholesterol deposits in the skin
     answer: d

19. The inherited disease called familial hypercholesterolemia appears to be due to
     a. defective low-density lipoprotein receptors
     b. the lack of a key enzyme involved in the production of high-density lipoprotein
     c. over-production of low-density lipoprotein particles
     d. the absence of an apoprotein that is a component of very low-density lipoproteins
     answer: a

20. An example of bile acids is
     a. taurine
     b. cholesteric acid
     c. cholic acid
     d. glycine
     e. oleic acid
     answer: c

21. The primary physiological function of bile acids is to
    a. emulsify dietary lipids to aid in their digestion
    b. provide a means of secreting steroid hormones
    c. activate lipases secreted into the small intestine
    d. raise the pH of the contents of the gallbladder
    answer: a

22. Which of the following structural features of cholic acid is not correct?
    a. it has 24 carbon atoms and no double bonds
    b. it has three hydroxyl groups, one each on rings A, B, and C
    c. its carboxyl group is at the terminus of a five-carbon side chain attached to the D-ring
    d. its hydroxyl group at the 3 position has, as it did in its parent molecule, cholesterol
    answer: d

23. Bile salts are effective in emulsifying triacylglycerol because they
    a. disperse triacylglycerol molecules so that they are in true solution
    b. form micelles with the polar faces of the cholate molecules on the outside and triacylglycerol molecules on the inside
    c. complex the cholesterol in the dietary lipids, leaving the triacylglycerol molecules more accessible to the aqueous phase
    d. bind to the triacylglycerol molecules forming a micelle that has the triacylglycerol on the outside where it is accessible to lipases
    answer: b

24. The steroid product formed in the mitochondria of adrenal glands and gonads by cleavage of the side chain of cholesterol is
    a. progesterone
    b. testosterone
    c. pregnenolone
    d. cortisol
    answer: c

25. Unlike the other steroid hormones, the hormone responsible for sexual characteristics in females has
    a. hydroxyl groups
    b. an aromatic ring
    c. no side chain at the 17-position
    d. more than 18 carbons
    answer: b

26. Which of the following is not a useful metabolic function of cholesterol?
     a. cholesterol serves as a structural component in membranes of animals
     b. in animals, all of the steroid hormones are derived from cholesterol
     c. energy may be obtained from the metabolism of pregnenolone, a metabolite of cholesterol
     d. cholesterol is a precursor of bile salts
     answer: c

27. During the formation of glycocholate, cholic acid, CoASH, and ATP form cholyl-CoA, AMP, and pyrophosphate. What reaction intermediate would you expect in this process?
     a. cholyl phosphate
     b. cholyl adenylate
     c. cholyl pyrophosphate
     d. cholyl group bound to a protein by a thioester linkage
     answer: b

28. During the conversion of cholesterol to cholic acid, one of the changes is the geometry of the hydroxy on carbon-4 of cholesterol. Provide a reason why nature made this geometry change.

     answer: This geometry change places the hydroxyl groups on the same "side" of cholic acid (or cholyl derivative). This geometry change produces hydrophilic and hydrophobic sides and allows cholyl compounds to function as a detergent.

Chapter 24 Amino Acid Biosynthesis and Nitrogen Fixation in Plants

1. In plants and microorganisms, the carbon skeletons of amino acids are derived from
   a. the amino acids of degraded proteins
   b. intermediates in the central metabolic pathways, such as the citric acid cycle and glycolysis
   c. the degradation of branched and long chain fatty acids
   d. directly from the fixation of carbon dioxide
   e. none of the above
   answer: b

2. Glycerate-3-phosphate is the central metabolic pathway intermediate used to produce the three members of this family of amino acids.
   a. glutamate family
   b. serine family
   c. aspartate family
   d. pyruvate family
   e. aromatic family
   answer: b

3. This amino acid family uses an intermediate of the citric acid cycle and methionine is one of its members
   a. glutamate family
   b. serine family
   c. aspartate family
   d. pyruvate family
   e. aromatic family
   answer: c

4. Chorismate, which is derived from a central metabolic pathway intermediate is a precursor of all of this family's members
   a. glutamate family
   b. serine family
   c. aspartate family
   d. pyruvate family
   e. aromatic family
   answer: e

5. α-Ketoglutarate is the common intermediate for members of this family of amino acids
   a. glutamate family
   b. serine family
   c. aspartate family
   d. pyruvate family
   e. aromatic family
   answer: a

6. This amino acid family includes ornithine, a non-protein amino acid which is converted to arginine
   a. glutamate family
   b. serine family
   c. aspartate family
   d. pyruvate family
   e. aromatic family
   answer: a

7. This family of amino acids includes valine, leucine and alanine
   a. glutamate family
   b. serine family
   c. aspartate family
   d. pyruvate family
   e. aromatic family
   answer: d

8. An important method used to determine the reaction pathways for the synthesis of the amino acids by *E. coli* involved the production of suitable auxotrophs which were
   a. strains that could grow on media containing ammonium salts as the only source of nitrogen
   b. mutants that required light for optimal growth
   c. E. coli strains that could not grow unless all of the common amino acids were included in the media
   d. mutants that could not grow on media containing ammonium salts as the only source of nitrogen, but could grow when the media was supplemented with one or more specific amino acids
   e. strains that could grow on media containing only amino acids and no glucose
   answer: d

9. In the synthesis of most of the common amino acids, there amino groups are derived directly or indirectly from
   a. ornithine
   b. aspartic acid
   c. glutamate
   d. glutamine
   answer: c

10. Although the enzyme, glutamate dehydrogenase , may be present in plants, L-glutamate appears to be synthesized by which of the following reaction?
    a. alanine + α-ketoglutarate $\rightarrow$ L-glutamate + pyruvate
    b. L-glutamine $\rightarrow$ L-glutamate + NH_3
    c. glutamine + NADPH + H^+ + α-ketoglutarate $\rightarrow$ 2 L-glutamate + $NADP^+$
    d. α-ketoglutarate + NH_4^+ + NADPH + H^+ $\rightarrow$ L-glutamate + $NADP^+$ + H_2O
    answer: c

11. The correct reaction for the synthesis of glutamine in plants and microorganisms is
    a. glutamate + NH_3 $\rightarrow$ glutamine + H^+
    b. 2 glutamate + NAD^+ $\rightarrow$ glutamine + H^+ + α-ketoglutarate + NADPH + H^+
    c. glutamate + NH_3 + NADPH + H^+ $\rightarrow$ glutamine + $NADP^+$ + H_2O
    d. glutamate + ATP + NH_4^+ $\rightarrow$ glutamine + ADP + Pi + H^+
    e. glutamate + alanine $\rightarrow$ glutamine + pyruvate
    answer: d

12. In *E. coli* glutamine synthase, an allosteric enzyme, is regulated
    a. negatively by ATP and positively by ADP
    b. Positively by AMP and negatively by about 15 of the common amino acids
    c. negatively by about eight nitrogenous compounds, including glucosamine-6-phosphate, carbamyl phosphate, several amino acids, and two nucleotides
    d. positively by glutamate and ammonium ions and negatively by arginine, asparagine, and ornithine
    answer: c

13. In addition to allosteric effectors, the activity of glutamine synthase in *E. coli* is regulated by
    a. phosphorylation of 12 serine residues that result in inactivation of the enzyme
    b. phosphorylation of 12 tyrosine residues that result in inactivation of the enzyme
    c. phosphorylation of 12 tyrosine residues that produce an activation of the enzyme
    d. association and dissociation of the 12 subunits of the enzyme, with α-ketoglutarate causing association and glutamine causing dissociation
    answer: b

14. In *E. coli* the activity of glutamine synthase is responsive to
    a. glutamine and α-ketoglutarate concentrations which effects the reversible adenylation an uridylation of enzymes that form an enzyme cascade
    b. glutamine and α-ketoglutarate concentrations that effects adenyltransferase, which in turn is affected by an enzyme kinase and phosphorylase
    c. glutamate concentration that effects several enzymes in an enzyme cascade involving adenylation and phosphorylation of enzymes
    d. several allosteric effectors including ammonium (+), glutamine (-), α-ketoglutarate (+), and glutamate (+)
    answer: a

15. Which of the following statements about the enzyme responsible for converting N_2 to NH_3 is false?
    a. molybdenum is present in the active enzyme
    b. the enzyme has several iron-sulfur centers
    c. in most bacteria, the enzyme consists of two identical subunits
    d. electrons generated by cell metabolism are transferred to Fe in the enzyme by ferrodoxin or flavodoxin
    answer: c

16. In animals the only source of sulfur for the synthesis of cysteine is
    a. H_2S
    b. the essential amino acid methionine
    c. O-acetyl-L-homoserine
    d. APS (adenosine-5'-phosphosulfate)
    answer: b

17. In the synthesis of cysteine by various plants and microorganisms, the reaction that incorporates sulfur (sulfide oxidation state) uses which of the following substrates?
    a. either o-acetyl-L-serine or O-acetyl-L-homoserine
    b. methionine
    c. L,L-cystathionine
    d. either L-serine or L-homoserine
    answer: a

18. The four final reactions in the synthesis of valine and isoleucine involve
    a. parallel reaction types but enzymes specific for each amino acid's intermediates
    b. only four enzymes that are capable of using alternative intermediates for the synthesis of both amino acids
    c. the same reaction types but the orders of the types are different
    d. a condensation reaction followed by oxidation, dehydration, and transamination reactions
    e. occur in different subcellular compartments
    answer: b

19. In the synthesis of valine and isoleucine, a two carbon moiety is condensed with either pyruvate or α-ketobutyrate. The cofactor to which the two-carbon unit is attached is
    a. coenzyme A
    b. thiamine pyrophosphate
    c. tetrahydrofolate
    d. lipoic acid
    e. AMP
    answer: b

20. The reason the herbacide, glyphosate, kills plants but is not toxic to mammals is that glyphosate specifically inhibits an enzyme involved in the synthesis of aromatic amino acids, but in mammals
    a. liver tissue degrades glyphosate rapidly
    b. an analogous enzyme does not exist
    c. glyphosate is rapidly conjugated and excreted in the urine
    d. a diet containing adequate quantities of the aromatic amino acids will compensated for the inhibition of aromatic amino acid biosynthesis
    answer: b

21. Although the enzyme activities for the five enzymes involved in the synthesis of tryptophan from chorismate are similar for all microorganism studied the activities are distributed differently on polypeptide chains. Which of the following arrangements is not observed among the species studied?
    a. two polypeptide chains associated to form an active enzyme
    b. two consecutive enzyme activities are located within a single polypeptide chain
    c. four of the enzyme activities are only present when a complex of all polypeptide chains is formed
    d. three enzymes activities are present when two polypeptide chains associate
    answer: c

22. Anthranilate synthase catalyzes a multi-step reaction with chorismate as a substrate. in this reaction an -NH_2 group is added, and it is derived from
    a. glutamate or ammonia
    b. glutamine or asparagine
    c. asparagine, exclusively
    d. glutamine or ammonia
    e. alanine
    answer: d

23. Which of the following statements about D-amino acids is false?
    a. many peptide antibiotics contain D-amino acids
    b. microbial cell walls frequently contain D-amino acids
    c. racemases that convert L-amino acids to the D-forms utilize Coenzyme A
    d. racemases may act on free amino acids or on residues that have been incorporated into peptides
    answer: c

24. Most nonprotein amino acids are derived
    a. by pathways unique for their synthesis
    b. from protein amino acids and their synthetic pathways
    c. from nutrients taken up by the organism
    d. by the complete synthesis of the carbon skeleton followed by a transamination reaction
    answer: b

25. Although only one amino acid is synthesized by the histidine pathway, the pathway is connected to another pathway responsible for the synthesis of
    a. pyrimidine nucleotides
    b. coenzyme A
    c. flavonoids
    d. Purine nucleotides
    e. heme
    answer: d

26. In the biosynthesis of proline from glutamate the γ-carboxyl is initially phosphorylated using an ATP followed by reduction of the mixed anhydride to produce a transient "hemiacetal-like" which eliminates a phosphate to produce L-glutamate semialdehyde. What was the function of the ATP in this process?

    answer: The mixed anhydride is more readily reduced and the phosphoryl group) from the ATP leaves as a phosphate which is a much better leaving group than the oxide or hydroxide ion that would be required without the phosphate.

27. Why does nature use an N-acetyl group on several of the intermediates in the arginine biosynthetic pathway?

    answer: The acetyl group allows that separation of products of glutamate that are in route to arginine from those that will become proline. Also, the N-acetyl group prevents the spontaneous cyclization of the γ-glutamate semialdehyde (which happens in the proline biosynthetic pathway).

28. L-Serine is produced from 3-phospho-D-glycerate by the sequential action of a dehydrogenase, transaminase, and phosphatase. What is the source of the alcohol oxygen in serine?

    answer: from water during the action of the phosphatase

29. L-Serine is produced from 3-phospho-D-glycerate by the sequential action of a dehydrogenase, transaminase, and phosphatase. What is (are) the source(s) of the two hydrogens on carbon three of serine?

    answer: from carbon three of 3-phospho-L-glycerate

30. L-Serine is produced from 3-phospho-D-glycerate by the sequential action of a dehydrogenase, transaminase, and phosphatase. How is it possible to produce an L-serine when the pathway starts with metabolite with the D-geometry?

    answer: During the reduction step carbon two becomes a non-chiral sp^2 carbon, which upon transamination was converted to the L-geometry.

Chapter 25 Amino Acid Metabolism in Vertebrates

1. Most vertebrates can synthesize approximately what fraction of the 20 amino acids found in proteins?
   a. 0.20
   b. 0.25
   c. 0.33
   d. 0.50
   e. 0.75
   answer: d

2. During evolution, higher animals have lost some of the genes needed for synthesizing the amino acids required for protein synthesis. Which of the following reasons is the most plausible?
   a. Higher animals have so many more genes than microorganisms that the number had to be reduced to keep the amount of cellular DNA at a manageable level.
   b. Some of the intermediates in the amino acid synthetic pathways are toxic to higher animals.
   c. Because higher animals eat other organisms that contain all of the amino acids, the amino acid biosynthetic pathways generally are superfluous.
   d. The genes for the enzymes involved in the de novo synthesis of amino acids are particularly labile, and if required for higher forms of life these species would quickly become extinct.
   e. None of the antecedents of higher animals had the required genes and higher animals were never successful in evolving the required genes.
   answer: c

3. The criteria for determining whether a particular amino acid is "essential" for humans is to supply adequate quantities of all of the other 19 amino acids:
   a. in the diets of test subjects and determine if more nitrogen is excreted than is consumed by the subjects
   b. in the diets of test subjects, and determine if there is a body weight loss by the subjects
   c. in media for culturing human cells, and determine if the cell count remains constant or decreases
   d. in the diets of human juveniles and determine if the subjects' growth is arrested
   answer: a

4. Although humans lack the enzymes necessary to synthesize the aromatic ring of tyrosine, that amino acid is not considered to be essential because
   a. phenylalanine can replace tyrosine in polypeptides
   b. there is an abundance of tyrosine in every natural diet imaginable
   c. the estrogen steroids contain an aromatic ring that can be utilized to synthesize tyrosine
   d. diets that contain an adequate quantity of phenylalanine, tyrosine can be synthesized by oxidation of phenylalanine
   answer: d

5. An exopeptidase is an enzyme that:
   a. catalyzes the hydrolysis of peptide bonds between either N-terminal or C-terminal amino acid residues and the remainder of the peptide chains
   b. is an enzyme that is produced and secreted only by the pancreas for the purpose of hydrolyzing peptides in the small intestine
   c. catalyzes the hydrolysis of peptide bonds between two amino acid residues, neither of which are at the chain's termini
   d. is a proteolytic enzyme produced and excreted for action outside the cell that synthesized the enzyme
   answer: a

6. When in excess, amino acids may be degraded, ant the first step in degradation of most amino acids is
   a. decarboxylation
   b. oxidation of the side chain
   c. elimination of functional groups such as hydroxyl and sulfhydryl groups
   d. transamination
   e. racemization
   answer: d

7. Most transaminases
   a. have high specificity for a particular amino acid and use pyridoxal phosphate as a cofactor
   b. use α-ketoglutarate as the amino group acceptor and can act on several different amino acids
   c. use NAD^+ to oxidize the amino group before transfer to a suitable acceptor
   d. form a multienzyme complex with glutamate dehydrogenase
   answer: b

8. The most likely explanation for the glutamate dehydrogenase involved in the degradation of amino acids being located in mitochondria is that
   a. an abundant supply of ATP is available for the degradation reactions
   b. the ammonia produced by the reaction can be incorporated into urea by the urea cycle which is present in mitochondria
   c. NADPH is required and it may be formed readily from the NADH generated by the citric acid cycle
   d. The NADH and α-ketoglutarate preoduced can be used by the mitochondrial electron transport system and TCA cycle directly
   answer: d

9. The principal compound formed to detoxify and excrete ammonia by birds and terrestrial reptiles is
   a. uric acid
   b. urea
   c. hippuric acid
   d. glutamine
   e. ammonium carbonate
   answer: a

10. The principal compound formed to detoxify and excrete ammonia by mammals is
    a. uric acid
    b. urea
    c. hippuric acid
    d. glutamine
    e. ammonium carbonate
    answer: b

11. The urea cycle occurs in
    a. the cytosol and mitochondria of liver tissue
    b. the chloroplasts of higher plants
    c. the mitochondria of kidney tissue
    d. the cytosol of most procaryotes
    e. the mitochondria of brain tissue
    answer: a

12. The urea and citric acid cycles are linked. The fumarate produced by the urea cycle is converted to oxaloacetate by the citric acid cycle, but the oxaloacetate cn be re-directed back to the urea cycle by
    a. condensing with acetyl-CoA and regenerating fumarate
    b. condensing with acetyl-CoA and forming α-ketoglutarate, which can be transaminated to form glutamate
    c. undergoing transamination to form aspartate
    d. undergoing decarboxylation and forming phosphoenolpyruvate
    answer: c

13. Ammonia is transported from skeletal muscle to the liver as
    a. ammonium salts
    b. glutamine and alanine
    c. glutamate and aspartate
    d. urea
    e. uric acid
    answer: b

14. The glycogenic amino acids are the ones that when degraded yield
    a. acetyl-CoA
    b. dicarboxylic acids that are intermediates in the citric acid cycle
    c. pyruvate
    d. products that can be converted to acetoacetyl-CoA
    answer: b

15. The ketogenic amino acids are the ones whose degradation yields compounds that
    a. can be converted to glucose
    b. are intermediates of the citric acid cycle
    c. are intermediates in glycolysis
    d. can be used to synthesize fatty acids
    answer: d

16. Most human genetic diseases that that involve amino acid metabolism are caused by a defective
    a. membrane transporter that prevent the absorption of a particular amino acid from the digested food
    b. enzyme for one step in the synthesis of a non-essential amino acid
    c. enzyme in the catabolism of a particular amino acid
    d. enzyme that is supposed to hydrolyze dietary proteins
    answer: c

17. In many of the human genetic diseases associated with amino acid metabolism
    a. an intermediate in the catabolic pathway of a particular amino acid accumulates and its high concentration affects the function of certain tissues
    b. protein synthesis is inhibited because of a lack of one of the required 20 amino acids
    c. many metabolic enzymes are defective because of erroneous substitutions caused by a lack of a particular amino acid
    d. key compounds, such as porphyrins and glutathione, are not synthesized since in catabolic processes intermediates beyond a defective enzyme are not produced
    answer: a

18. Which of the following is not a characteristic of human genetic diseases associated with amino acid metabolism?
    a. the affected fetus usually develops normally
    b. symptoms begin to appear with puberty
    c. mental retardation occurs in many of the diseases
    d. for many of the diseases, unusual products are excreted in the urine
    answer: b

19. Which of the following is not a product synthesized from amino acids by humans?
    a. glutathione
    b. a biologically active amine
    c. biotin
    d. porphyrin
    answer: c

20. Which of the following statements about δ-aminolevulinate is false?
    a. it is a dicarboxylic acid that is an intermediate in the synthesis of porphyrins
    b. it is synthesized from glycine and succinyl-CoA by mammals and some bacteria
    c. it is synthesized from glutamate by plants and some bacteria
    d. it contains five carbon atoms, a keto group, and one amino group
    answer: a

21. Glutathione is
    a. γ-cysteinlyglutamylglycine
    b. synthesized by activation of the amino acids it contains with the tRNAs
    c. only present in red blood cells of mammals where it maintains the appropriate redox state of iron in hemoglobin
    d. used physiologically to help maintain the sulfhydryl groups of proteins in a reduced state
    answer: d

22. In the γ-glutamyl cycle
    a. 5-oxoproline is an intermediate in the conversion of L-glutamate to γ-glutamate
    b. γ-glutamylcsteine synthase uses the following substrates: γ-glutamate, cysteine, and ATP
    c. the γ-glutamyl residue is transferred from glutathione to an amino acid acceptor outside cells that contain the other enzymes of the cycle
    d. the dipeptide, cysteinylglycine, is condensed with γ-glutamylphosphate
    answer: c

23. Which of the following includes only components associated with the urea cycle?
   a. citrulline, NADPH, carbanoyl phosphate
   b. ammonia, uric acid, aspartate, fumarate
   c. AMP, argininosuccinate, PPi, citrate
   d. fumarate, Pi, ADP, bicarbonate
   e. ornithine, aspartate, asparagine, arginine
   answer: d

24. How many molecules of δ-aminolevulinate are required to make one molecule of porphyrin?
   a. 2
   b. 4
   c. 6
   d. 8
   e. none of these
   answer: d

25. The γ-glutamyl cycle is
   a. involved in the synthesis of nitrogen waste products
   b. involved in amino acid transport
   c. has a redox role
   d. is an exercise machine
   e. none of the above
   answer: b

26. Would you expect to observe as many genetic diseases in chickens as there have been found in humans?

   answer: Fewer are anticipated in chickens because the growing embryo is "on it own" in the egg. In mammals, the mother can in many cases complement the embryo metabolically (e.g., disposing of potentially lethal concentrations of a metabolite that occurs before a faulty enzyme, and supplying the product of the pathway that contains the faulty enzyme.) This latter sequence could allow for the birth of an apparently normal individual.

27. What is very unusual about the involvement of ATP with carbamoyl phosphate synthetase?

   answer: It is essentially the only reaction that utilizes more than one ATP. This suggests that a fundamental biochemical principle is being violated (the energies of multiple ATP's are not combined to overcome an energy barrier). In reality the ATP's are used sequentially in the reaction mechanism to produce two separate bonds in carbamoyl phosphate.

28. What is the basic difference between transamination and deamination of amino acids?

answer: Transamination preserves the number of moles of amino acid but it does change the identity of the amino acid. Deamination involves a reduction of an amino acid and its conversion to an alpha keto acid and ammonia.

Chapter 26 Nucleotides

1. Which of the following statements about nucleotides is false?
   a. ATP is a nucleotide that is used by all organisms to transfer energy form energy-yielding reactions to energy-consuming reactions.
   b. All nucleotides serve as precursors for nucleic acids.
   c. All nucleotides consist of a nitrogen containing base, a pentose moiety, and at least one phosphoryl group.
   d. Some nucleotides regulate enzymes by serving as allosteric effectors, or as donors of groups for covalent modification.
   answer: b

2. A nucleoside consists of
   a. a purine or pyrimidine, a sugar moiety, and at least one phosphoryl group
   b. a nitrogen-containing nucleobase and a phosphoryl group
   c. two nitrogen-containing nucleobases joined by a pentose residue
   d. a nucleobase and a pentose residue
   answer: d

3. Nucleotides typically contain which of the following bond types?
   a. an α-N-glycosidic bond
   b. a β-N glycosidic bond
   c. an α-O-glycosidic bond
   d. a β-O glycosidic bond
   answer: b

4. Which of the following has the most aromatic character?
   a. uridine
   b. cytidine
   c. adenosine
   d. guanosine
   answer: c

5. Which of the following is not an acceptable name for AMP?
   a. adenosine-5'-monophosphate
   b. adenylate
   c. adenine mononucleotide phosphate
   d. adenosine-5'-phosphate
   answer: c

6. The nucleotide, dGTP, consists of
    a. guanine, ribose, and three phosphoryl groups
    b. GMP pluss two phosphoryl groups
    c. guanine, 2-deoxyribose, and three phosphoryl groups
    d. guanosine, 2-deoxyribose, and three phosphoryl groups
    answer: c

7. Which of the following statements correctly describes the chemical state of the three phosphoryl groups in NTPs *in vivo*?
    a. A majority of the molecules will have a charge of -4, and may form salts with Mg^{2+} or Ca^{2+}
    b. A majority of the molecules will have a charge of -3, and may form salts with Mg^{2+} or Ca^{2+}
    c. A majority of the molecules will have a charge of -4, and may form salts with Mg^{2+}
    d. A majority of the molecules will have a charge of -2, and may form salts with Mg^{2+} or Ca^{2+}
    answer: a

8. For the *denovo* synthesis of purine nucleotides, the important precursor, phosphoribosylpyrophosphate, (PRPP) is synthesized by which of the following reactions?
    a. α-D-ribose + 3 ATP $\rightarrow$ PRPP + 3 ADP
    b. α-D-ribose-1-phosphate + 2 ATP $\rightarrow$ PRPP + 2 ADP
    c. α-D-ribose-5-phosphate + 2 ATP $\rightarrow$ PRPP + 2 ADP
    d. α-D-ribose-5-phosphate + ATP $\rightarrow$ PRPP + AMP
    answer: d

9. In the *de novo* synthesis of purines the nitrogen atoms in the purine bases are derived from
    a. arginine, proline, aspartate, and glutamine
    b. aspartate, alanine, and two molecules of glutamate
    c. aspartate, glycine, and two molecules of glutamine
    d. glutamate, glutamine, asparagine, and proline
    answer: c

10. In the *de novo* synthesis of purines the carbon atoms in the purine base are derived from
    a. alanine, 2 CO_2, and 10-formyltetrahydrofolate
    b. glycine, CO_2, and two molecules of 10-formyltetrahydrofolate
    c. aspartate, CO_2 (via biotin), and 10-formyltetrahydrofolate
    d. glycine, fumarate, and CO_2
    answer: b

11. In the *de novo* synthesis of purines the first purine nucleotide formed is
    a. inosine monophosphate
    b. AMP
    c. xanthosine monophosphate
    d. guanylate
    e. UMP
    answer: a

12. The primary substrates for the synthesis of purine and pyrimidine nucleotides by the so called "salvage" pathways are
    a. 5-phospho-α-D-ribosyl-1-pyrophosphate and ATP
    b. nucleobases or nucleosides
    c. IMP and XMP
    d. AMP, GMP, UMP, dTTP, and CMP
    answer: b

13. In the *de novo* synthesis of pyrimidine nucelotides the carbon atoms of the pyrimidine ring are derived from
    a. bicarbonate and aspartate
    b. glycine and carbon dioxide
    c. glutamine, 10-formyltetrahydrofolate, and glycine
    d. asparagine, carbon dioxide and glycine
    answer: a

14. The carboxyl group released as carbon dioxide when orotidine-5'-monophosphate (OMP) is converted to UMP, was derived from
    a. bicarbonate
    b. carbon dioxide
    c. 10-formyltetrahydrofolate
    d. aspartate
    answer: d

15. The riboncleotide reductase that catalyzed the conversion of a ribose moiety to a 2-deoxyribose moiety is specific for
    a. ADP, GDP, CDP, TDP, and UDP
    b. ADP, GDP, CDP and UDP
    c. AMP, GMP, CMP, TMP, and UMP
    d. IMP and OMP
    answer: b

16. The proper substrate for thymidylate synthase, which adds a methyl group to a uracil moiety in a nucleotide to form a thymine moiety, is
    a. dUMP
    b. UMP
    c. dUDP
    d. UTP
    answer: b

17. The methyl group donor in the reaction catalyzed by thymidylate synthase is
    a. 5'-deoxyadenosylcobalamin
    b. 5, 10-methylenetetrahydrofolate
    c. S-adenosylmethionine
    d. 10-formyl-7,8-dihydrofolate
    answer: b

18. An important enzyme for the conversion of purine bases to nucleotides is hypoxanthine-guanine phosphoribosyltransferase which catalyzes the formation of
    a. HMP or GMP from hypoxanthine or guanine plus PRPP
    b. HMP or GMP from hypoxanthine or guanine plus α-D-ribosyl-1-pyrophosphate
    c. IMP or GMP from hypoxanthine or guanine plus 5-phospho-α-D-ribosyl-1-pyrophosphate
    d. AMP or GMP from hypoxanthine or guanine plus α-D-ribosyl-1-pyrophosphate
    answer: c

19. People who lack hypoxanthine-guanine phosphoribosyltransferase exhibit
    a. very high blood levels of orotic acid
    b. elevated intercellular pools of AMP and GMP
    c. very low intercellular levels of PRPP and IMP
    d. very high rates of purine biosynthesis
    answer: d

20. Which of the following reactions is typical of a purine nucleoside kinase?
    a. adenosine + ATP → AMP + ADP
    b. cytosine + ATP → CMP + ADP
    c. adenosine + ATP → 2 AMP + Pi
    d. guanine + ATP → AMP + GMP + Pi
    answer: a

21. Before absorption, digestive enzymes degrade nucleic acids to
    a. nucleotides
    b. oligonucleotides
    c. nucleobases
    d. nucleosides and nucleobases
    answer: d

22. People suffering from gout exhibit
    a. elevated plasma levels of xanthine and kidney stones of the same material
    b. elevated plasma levels of uric acid and deposits of monosodium urate in bone joints
    c. arrested growth in certain tissues, particularly in the growth zones of bones in the feet
    d. high urinary excretion of allatoin and urea
    answer: b

23. The catabolism of the cytidine, 2-deoxycytidine, and uracil pyrimidine rings produces
    a. β-alanine
    b. β-aminoisobutyric acid
    c. ammonia and carbon dioxide only
    d. urea and carbon dioxide
    answer: a

24. Although several different nucleotides are involved in the regulation of the synthesis of 2-deoxyribonucleotides, a key enzyme appears to be
    a. thymidylatesyntase
    b. thioredoxin reductase
    c. carbamoyl aspartate synthase
    d. ribonucleotide reductase
    answer: d

25. Describe the structure of PRPP.

    answer: PRPP is a ribose with a phosphate moiety on carbon five and a pyrophosphate moiety in the α-geometry on carbon one.

26. Describe the conventions used in the nucleotide abbreviations.

    answer; The general formula, #'-dNMP, is followed where the # indicates the carbon number on the sugar that the phosphoryl moietie(s) is (are) located with no number in this position indicating carbon five, the ' indicates a sugar carbon, a lower case d indicates a 2-deoxyribose, no letter in this position indicates a ribose, N identifies the specific base (e.g., A, T, G, G, U), M indicates the number of phosphoryl moieties (M = mono, D = di, T = tri), and P stands for phosphate.

27. In twenty words or less indicate how the sulfa drugs function.

    answer: Sulfa drugs compete with p-aminobenzoate during the biosynthesis of folic acid. Folic acid is not synthesized by humans.

Chapter 27 Integration of Metabolism in Vertebrates

1. More than 90% of the fuel reserves of an average human are in the form of
   a. phospholipids.
   b. triacylglycerides.
   c. glycogen.
   d. proteins.
   answer: b

2. The brain accounts for about _____% of the glucose consumption in a resting human.
   a. 10
   b. 25
   c. 40
   d. >50
   answer: d

3. Which of the following is a major carbohydrate storage reserve in humans?
   a. the heart
   b. adipose tissue
   c. the liver
   d. the brain
   answer: c

4. The Cori cycle converts lactate (generated during anaerobic glycolysis) back to glucose. This occurs
   a. in the bloodstream.
   b. in the muscle cells.
   c. only in the liver.
   d. both in the liver and in heart muscle.
   answer: d

5. The substance preferentially oxidized by heart muscle is
   a. glucose
   b. glycogen
   c. fatty acids
   d. triacylglycerides
   answer: c

6. The location of epinephrine synthesis is
   a. the β cells of the pancreas.
   b. the thyroid.
   c. the adrenal medula.
   d. the skin.
   answer: c

7. Of the following organs or tissues, which **cannot catabolize fatty acids directly?**
   a. the brain
   b. the heart
   c. skeletal muscle
   d. the liver
   answer: a

8. After prolonged starvation, the most abundant energy supply molecule(s) in the blood stream is (are)
   a. glucose.
   b. fatty acids.
   c. ketone bodies.
   d. ATP.
   answer: c

9. In adipose tissue, insulin stimulates
   a. glycogen breakdown.
   b. glycogen synthesis.
   c. triacylglycerol breakdown.
   d. triacylglycerol synthesis.
   answer: c

10. Which of the following is NOT a peptide hormone?
   a. insulin
   b. glucagon
   c. melatonin
   d. somatostatin
   answer: c

11. The function of glucagon is to
   a. stimulate glycogen breakdown.
   b. stimulate glycogen synthesis and release of lipid.
   c. stimulate insulin release.
   d. inhibit somatotropin release.
   answer: b

12. Some people with kidney disease have a form of Vitamin D-resistant Rickets that results from their inability to convert vitamin D_3 to an active hormone. These individuals are unable to make
   a. 7 dehydrocholesterol.
   b. 25-hydroxycholesterol.
   c. 24, 25-dihydroxycholesterol.
   d. 1, 25-dihydroxycholesterol.
   answer: d

13. Epinephrine, norepinephrine and dopamine are all related in that they are all derived from
    a. serine.
    b. glutamine.
    c. tyrosine.
    d. tryptophan.
    answer: c

14. Testosterone belongs to a class of hormones called
    a. androgens.
    b. glucocorticoids.
    c. mineralcroticoids.
    d. progestins.
    answer: a

15. Hormone receptors that activate or inhibit adenyl cyclase must first activate
    a. protein kinase C.
    b. a G protein.
    c. phospholipase C.
    d. protein kinase A.
    answer: b

16. cAMP is a second messenger that activates
    a. protein kinase A.
    b. the insulin receptor protein.
    c. G proteins.
    d. phospholipase C.
    answer: a

17. In the absence of a hormone signal, the α subunit of a G protein binds to
    a. cAMP
    b. Ca^{2+}
    c. GDP
    d. GTP
    answer: c

18. cGMP synthesis is often stimulated by
    a. cAMP
    b. nitric oxide (NO)
    c. Ca^{2+}
    d. inositol trisphosphate (IP_3)
    answer: b

19. The amino acid precursor of nitric oxide is
   a. arginine
   b. asparagine
   c. glutamine
   d. tyrosine
   answer: a

20. Steroid hormones generally act through
   a. G proteins.
   b. DNA transcription regulation.
   c. phospholipase C.
   d. calmodulin.
   answer: b

21. Chronic stimulation by hormones such as insulin leads to a reduction in the number of receptor proteins on the surface of target cell membranes. This occurs through a mechanism that involves
   a. endocytosis of vesicles that engulf the receptors, internalize them and eventually transport them for degradation in lysosomes.
   b. uncoupling of the second messenger activation by hormone-bound receptor proteins.
   c. feedback inhibition of protein synthesis in the target cells.
   d. release of excess receptor proteins into the bloodstream.
   answer: a

22. Ras acts to stimulate cell proliferation
   a. through inhibition of growth factor receptor tyrosine kinases.
   b. through stimulation of growth factor receptor tyrosine kinases.
   c. through activation of a serine-threonine kinase cascade.
   d. through binding to transcription promoter sites on DNA.
   answer: c

23. Preproinsulin contains a hydrophobic signal sequence that does not exist in the active hormone. The function of this hydrophobic sequence is to
   a. insure correct folding of insulin into its active form.
   b. target the protein for transport across the endoplasmic reticular membrane.
   c. inhibit hormone activity until it is secreted into the blood stream.
   d. direct cell proteases in correctly processing proinsulin to form insulin.
   answer: b

24. Steroid hormone receptors
    a. activate G proteins.
    b. bind to DNA to either activate or inhibit transcription.
    c. interact with IP$_3$ to stimulate the release of intercellular Ca^{2+} stores.
    d. function as tyrosine kinases when bound to their respective hormones.
    answer: b

25. Cleavage of phosphadidylinositol-4,5-bisphosphate (PIP$_2$) results in a transient increase in intercellular Ca^{2+} through the intermediary action of
    a. calmodulin.
    b. diacylglycerol.
    c. IP$_3$ (inositol trisphosphate).
    d. protein kinase C.
    answer: c

26. Pseudohypoparathyroidism is an inherited disorder with symptoms of parathyroid hormone deficiency. However, when affected individuals are tested, they are found to overproduce PTH. In addition, these individuals exhibit abnormally low levels of urinary cAMP in response to exogenously added PTH. Which of the following genetic defects could you RULE OUT as a cause of this disorder?
    a. a defective PTH receptor protein.
    b. a defective G protein.
    c. defective adenyl cyclase.
    d. defective phospholipase C.
    answer: d

27. Preproinsulin contains a hydrophobic signal sequence that does not exist in the active hormone. The function of this hydrophobic sequence is to
    a. insure correct folding of insulin into its active form.
    b. target the protein for transport across the endoplasmic reticular membrane.
    c. inhibit hormone activity until it is secreted into the blood stream.
    d. direct cell proteases in correctly processing proinsulin to form insulin.
    answer: b

28. The effect of nitric oxide (NO) on the heart is to
    a. relax blood vessels, thereby increasing the blood flow to the heart muscle.
    b. release Ca^{2+} from intercellular stores, thus stimulating and strengthing heart muscle contraction.
    c. constrict blood vessels, thereby decreasing blood flow to the heart muscle.
    d. reduce intercellular Ca^{2+} from intercellular stores, thus stimulating and stimulating heart muscle contraction.
    answer: a

171

29. The chemical structure shown at the right is the plant hormone, auxin. From its structure, auxin appears to be

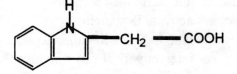

   a. a deamination product of tryptophan.
   b. a deamination product of tyrosine.
   c. a derivative a histamine.
   d. a metabolite of s-adenosyl methionine.
   answer: a

30. The precursor of the plant hormone, ethylene is
   a. auxin
   b. s-adenosylmethionine
   c. adenine
   d. pyruvate
   answer: b

Chapter 28 Neurotransmission

1. The Na⁺ and K⁺ concentrations across the membrane of a giant squid axon are:

    a.

	Na^+	K^+
inside:	50 mM	400 mM
outside:	440 mM	20 mM

    b.

	Na^+	K^+
inside:	440 mM	20 mM
outside:	50 mM	400 mM

    c.

	Na^+	K^+
inside:	440 mM	400 mM
outside:	50 mM	20 mM

    d.

	Na^+	K^+
inside:	50 mM	20 mM
outside:	440 mM	400 mM

    answer: a

2. The unequal distribution of ion concentrations across a nerve membrane is maintained by
    a. ion channels.
    b. neurotransmitters.
    c. neurotransmitter receptors.
    d. the Na^+, K^+-ATPase.
    answer: d

3. The resting potential across a squid axonal membrane is about
    a. 0 mV
    b. -60 mV
    c. +40 mV
    d. variable between -60 mV and +40 mV
    answer: b

4. The long range signaling component of a nerve cell is called the
   a. axon.
   b. dendrite.
   c. myelin sheath.
   d. cell body.
   answer: a

Questions 5 through 9 refer to the figure shown below:

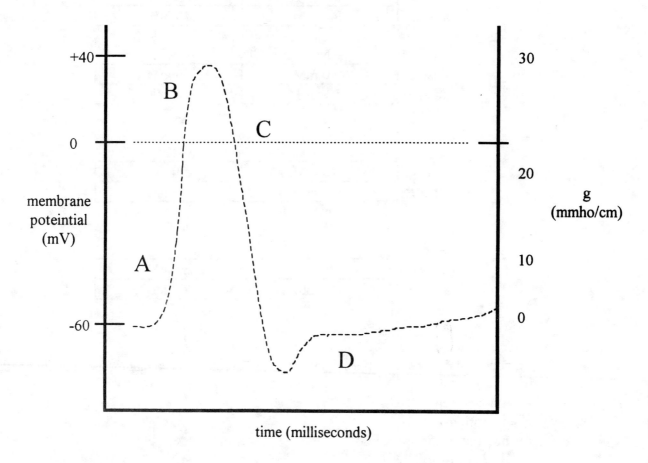

5. The time-dependent change in the membrane potential in the figure shown above is called
   a. the action potential.
   b. the resting potential.
   c. the threshold potential.
   d. hyperpolarization.
   answer: a

6. During the portion of the curve marked "B"
   a. only the K^+ channel is open.
   b. only the Na^+ channel is open.
   c. both the K^+ and Na^+ channels are open.
   d. neither the K^+ nor the Na^+ channels are open.
   answer: b

7. The threshold value for the initiation of the change in membrane potential in the figure shown above is
   a. 20 mV
   b. 40 mV
   c. 60 mV
   d. -60 mV
   answer: a

8. During the portion of the cycle marked "C"
   a. only the K^+ channel is open.
   b. only the Na^+ channel is open.
   c. both the K^+ and Na^+ channels are open.
   d. neither the K^+ nor the Na^+ channels are open.
   answer: b

9. The state indicated by the portion of the cycle marked "D" is called
   a. depolarization.
   b. repolarization.
   c. hyperpolarization.
   d. hypopolarization.
   answer: c

10. Saxitoxin is a potent nerve toxin made by marine dinoflagellates associated with the "red tide." It specifically blocks
    a. the Na^+, K^+-ATPase.
    b. the Na^+ channel.
    c. the K^+ channel.
    d. acetylcholinesterase.
    answer: b

11. Phosphorylation of Na^+ channels
    a. inhibits Na^+ influx.
    b. stimulates Na^+ influx.
    c. reduces the gating potential required for Na^+ influx.
    d. irreversibly inactivates the Na^+ channels.
    answer: a

12. The specific connection between nerve cells and target cells is called
    a. an axon.
    b. a dendrite.
    c. a synaptic vesicle.
    d. a synapse.
    answer: d

13. The function of acetylcholinesterase is to
    a. prevent the fusion of synaptic vesicles with the plasma membrane.
    b. degrade acetylcholine to acetate and choline in the synaptic cleft.
    c. add acetyl CoA to choline to form acetylcholine.
    d. stimulate release of acetylcholine from synaptic vesicles.
    answer: b

14. Release of acetylcholine from synaptic vesicles in the nerve cells is triggered by
    a. Ca^{2+} uptake.
    b. ATP hydrolysis
    c. neurotransmitter uptake.
    d. K^+ uptake.
    answer: a

15. Inhibitors of acetylcholinesterase, such as organophosphorus compounds, bind to an
    essential _____ residue at the active site of the enzyme.
    a. aspartate
    b. asparagine
    c. serine
    d. tyrosine
    answer: c

16. A regulatory mechanism that resembles a "ball and chain" has been proposed for
    a. the acetylcholine receptor protein.
    b. the K^+ channel.
    c. monoamine oxidase.
    d. acetlycholinesterase.
    answer: b

17. Parkinson's disease is a neurological disorder that is associated with
    a. overproduction of gama-aminobutyric acid.
    b. underproduction of monoamine oxidase.
    c. overproduction of dopamine.
    d. underproduction of dopamine.
    answer: d

18. Neurotransmitter receptors are located within
    a. gap junctions.
    b. presynaptic membranes.
    c. postsynaptic membranes.
    d. the synaptic cleft.
    answer: c

19. The amino acid precursor of neurotransmitters called catecholamines is
    a. serine.
    b. glycine.
    c. histamine.
    d. tyrosine.
    answer: d

20. Unlike rapid synaptic response, slow synaptic responses involve between the neurotransmitter receptor and a class of proteins called
    a. ion pumps.
    b. G proteins.
    c. ion channels.
    d. monoamine oxidases.
    answer: b

21. Which of the following statements is TRUE?
    a. Neurotransmitters bind to their receptor proteins on the same side of the membrane as the ion gate.
    b. Neurotransmitters bind to their receptor proteins on the exterior face of the receptor protein.
    c. Neurotransmitters bind to their receptor proteins within the ion pore.
    d. Neurotransmitters bind to their receptor proteins on the interior face of the receptor protein.
    answer: b

22. Storage of short term memory appears to involve
    a. activation of gene expression and synthesis of new proteins.
    b. inhibition of new protein synthesis.
    c. covalent modification of preexisting proteins.
    d. the growth of new synaptic connections.
    answer: c

23. Dopamine is derived from L-dopa through
    a. decarboxylation.
    b. amination.
    c. deamination.
    d. hydroxylation.
    answer: a

24. Dopamine receptor-blocking drugs are used to treat
    a. Parkinson's disease.
    b. Alzheimer's disease.
    c. poisoning by acetylcholinesterase inhibitors.
    d. psychological diseases such as schizophrenia.
    answer: d

25. Binding of acetylcholine to the acetlycholine receptor protein initiates an action potential in the recipient membrane through
    a. selectively increasing the inward flow of K^+ relative to the influx of Na^+.
    b. decreasing the ionic permeability of the postsynaptic membrane.
    c. inducing acetylcholine release from the postsynaptic cell.
    d. transiently opening an ion channel through which both Na^+ and K^+ can flow.
    answer: d

26. Exocytosis, fusion and release of acetylcholine in response to the arrival of an action potential in the presynaptic membrane occurs in
    a. less than 1 millisecond.
    b. about 5 milliseconds
    c. about 1 second.
    d. about 5 seconds.
    answer: a

27. Following the release of acetylcholine from a synaptic vesicle, there is a cycle that regenerates fresh vesicles, each packed with 10^3 to 10^4 molecules of acetylcholine. This cycle takes
    a. less than 1 millisecond.
    b. about 5 milliseconds.
    c. about 5 seconds.
    d. about 60 seconds.
       answer: d

28. For a hypothetical membrane that is selectively permeable to Na^+ only, calculate the transmembrane electrical potential given the following conditions:
$$R = 8.31451 \text{ J/K} \bullet \text{mole}$$
$$T = 298 \text{ K}$$
$$F = 96485 \text{ J/V} \bullet \text{mole}$$
$$[Na^+] \text{ inside} = 100 \text{ mM}$$
$$[Na^+] \text{ outside} = 440 \text{ mM}$$
    a. -38 mV
    b. +38 mV
    c. -56 mV
    d. +56 mV
    answer: b

178

Chapter 29 Vision

1. Rhodopsin is a member of a family of membrane receptor proteins that all have ____ transmembrane helical segments.
   a. two
   b. four
   c. seven
   d. ten
   answer: c

2. 11-cis-retinal is derived from
   a. vitamin A
   b. vitamin B_{12}
   c. vitamin C
   d. vitamin D
   answer: a

3. The cofactor in the active form of rhodopsin is linked to the protein
   a. by a protonated Schiff base linkage to an arginine residue.
   b. by a non-protonated Schiff base linkage to an arginine residue.
   c. by a protonated Schiff base linkage to a lysine residue.
   d. by a non-protonated Schiff base linkage to a lysine residue.
   answer: c

4. The outer segment of each rod cell contains
   a. 3 disks.
   b. between 10 and 40 disks.
   c. between 500 and 2000 disks.
   d. about 200,000 disks.
   answer: c

5. The lens of the eye is made of a transparent protein called
   a. crystallin
   b. opsin
   c. bathorhodopsin
   d. lumirhodopsin
   answer: a

6. In rod cells, rhodopsin absorbs light maximally at
   a. 220 nm
   b. 380 nm
   c. 500 nm
   d. 750 nm
   answer: c

7. Human cone cells are sensitive to color because
   a. light of different wavelengths has different energies which are more or less efficient in activating rhodopsin.
   b. human cone cells have filters that allow only light of certain wavelengths to penetrate their disc membranes.
   c. light rays entering the eye are refracted differently, depending on the wavelength and, because the cofactor of rhodopsin is oriented specifically within the disc membranes, rhodopsin activation is sensitive to the angle of refraction.
   d. cone cells have three different rhodopsin molecules, each of which is maximally sensitive to a different wavelength.
   answer: d

8. The orientation of the retinal with respect to the plane of the disc membrane is
   a. parallel.
   b. perpendicular.
   c. parallel when it is in the form of 11-cis-retinal, but perpendicular when it is all-trans-retinal.
   d. perpendicular when it is in the form of 11-cis-retinal, but parallel when it is all-trans-retinal.
   answer: a

9. When the electron density increases near the nitrogen atom of the Shiff base linkage in rhodopsin,
   a. the wavelength of the absorption maximum decreases.
   b. the wavelength of the absorption maximum increases.
   c. the wavelength of the absorption maximum remains the same.
   d. there are two absorption maxima in the region between 380 nm and 630 nm.
   answer: a

10. When opsin is mixed with retinal in vitro,
    a. neither 11-cis-retinal nor all-trans-retinal will bind.
    b. both 11-cis-retinal and all-trans-retinal will bind.
    c. only 11-cis-retinal will bind.
    d. only all-trans-retinal will bind.
    answer: c

11. Bathorhodopsin is the first metastable product of the photochemical reaction of rhodopsin. The cofactor in bathorhodopsin is
    a. 11-cis-retinal.
    b. all-trans-retinal.
    c. 9-cis-retinal.
    d. 13-cis-retinal.
    answer: b

12. Which of the following activates transducin?
    a. bathorhodopsin
    b. lumirhodopsin
    c. metarhodopsin I
    d. metarhodopsin II
    answer: d

13. Transducin is
    a. a Ca^{2+} binding protein.
    b. a phosphodiesterase.
    c. a Na^+ channel activator.
    d. a G protein.
    answer: d

14. The sequence of events following absorption of a photon of light by a rhodopsin
    molecule in a rod cell is as follows:
    a. 1) transducin is activated, 2) phosphodiesterase is activated, 3) cGMP is
       hydrolyzed, 4) Na^+ channels close.
    b. 1) transducin is activated, 2) cGMP is synthesized, 3) phosphodiesterase is
       inactivated, 4) Na^+ channels close.
    c. 1) transducin is activated, 2) phosphodiesterase is activated, 3) cGMP is
       hydrolyzed, 4) Na^+ channels open.
    d. 1) transducin is activated, 2) phosphodiesterase is activated, 3) cGMP is
       synthesized, 4) Na^+ channels open.
    answer: a

15. When rhodopsin is activated by light, the rod cell is
    a. depolarized by the transient opening of Na^+ ion channels.
    b. hyperpolarized by the transient closure of Na^+ ion channels.
    c. hypopolarized by the transient opening of Na^+ ion channels.
    d. polarized by the transient opening of Na^+ ion channels.
    answer: b

16. Of the following, which is a single subunit integral membrane protein with a molecular
    mass of about 38,000?
    a. transducin
    b. phosphodiesterase
    c. gyanlylate cyclase
    d. opsin
    answer: d

17. As rhodopsin cycles through excitation by light followed by regeneration,
    a. 11-cis-retinal is isomerized to all-trans-retinal which is then regenerated in-situ through a process that involves 9-cis-retinal and phosphatidylcholine.
    b. All-trans-retinal is isomerized to 11-cis-retinal, which is then released from the rhodopsin. The 11-cis-retinal is then transferred to the pigment epithelium where it is converted back to all-trans-retinal.
    c. 11-cis-retinal is isomerized to all-trans-retinal which is then released from rhodopsin. The all-trans-retinal is transferred to the pigment epithelium where it is converted back to 11-cis-retinal in a series of reactions that involve reduction by NADH, esterification, conversion to 11-cis-retinol, and finally oxidation to form 11-cis-retinal.
    d. All-trans-retinal is isomerized to 11-cis-retinal, which is then converted back to form all-trans-retinal in a reaction that requires light.
    answer: c

18. Guanlylate cyclase (the enzyme that converts GTP to cGMP) is strongly inhibited by
    a. arrestin.
    b. Na^+.
    c. Ca^{2+}.
    d. the α subunit of transducin.
    answer: c

19. Phosphodiesterase is activated by
    a. phosphorylation at multiple serine and threonine sites.
    b. a complex between the α subunit of transducin and GTP.
    c. Na^-.
    d. all-trans-retinal.
    answer: b

20. The electrochemical potential across the rod cell cytoplasmic membrane is generated by
    a. the Na^-, K^+-ATPase.
    b. the plasma membrane Ca^{2+}-ATPase.
    c. phosphodiesterase.
    d. movement of protons across the plasma membrane in response to light activation of rhodopsin.
    answer: a

21. The function of arrestin is to
    a. inhibit production of cGMP.
    b. phosphorylate rhodopsin.
    c. bind to phosphorylated rhodopsin, thus preventing continued activation of transducin.
    d. bind to phosphodiesterase, thus rendering it inactive.
    answer: c

22. Reversal of the activation of the α subunit of transducin is achieved by
    a. replacement of bound GDP by GTP.
    b. binding of cGMP.
    c. hydrolysis of bound GTP, leaving bound GDP and free inorganic phosphate.
    d. release of cGMP.
    answer: c

23. In the pigment epithelium, the energy required to convert all-trans-retinal to 11-cis-retinal is supplied by
    a. GTP hydrolysis.
    b. light.
    c. the Na^+ gradient across the cell membrane.
    d. hydrolysis of a phospholipid.
    answer: d

24. Each molecule of rhodopsin that is converted to metarhodopsin II causes activation of
    a. 1 molecule of transducin.
    b. about 500 molecules of transducin.
    c. about 25,000 molecules of transducin.
    d. about 2.5×10^6 molecules of transducin.
    answer: b

25. Guanlylate cyclase catalyzes the synthesis of cGMP from GTP. If you transiently inhibited this enzyme (for < 10 milliseconds),
    a. you would transiently arrest the ability of a rod cell to respond to light.
    b. the effect would be the same as if the rod cell were responding to a light signal.
    c. there would be no effect on the rod cell's ability to respond to light, however, there would be a negative effect on the ability to regenerate 11-cis-retinal.
    d. 11-cis-retinal would spontaneously isomerize to form all-trans-retinal even in the absence of light.
    answer: b

26. GppNp is a nonhydrolyzable analog of GTP that can bind to transducin in the presence of light-activated rhodopsin. Furthermore, the complex between the transducin α subunit and GppNp can bind to the regulatory peptide of phosphodiesterase. Thus, in the presence of GppNp, you would expect that
    a. rhodopsin, transducin and phosphodiesterase would all function normally.
    b. rhodopsin would not activate transducin.
    c. rhodopsin would activate transducin but, transducin would not activate phosphodiesterase.
    d. rhodopsin would activate transducin and transducin would activate phosphodiesterase but, the signal to activate transducin could not be reversed.
    answer: d

27. Transducin α–GDP binds to
    a. the regulatory peptide of phosphodiesterase.
    b. the catalytic subunit of phosphodiesterase.
    c. the β and γ subunits of transducin.
    d. guanlylate cyclase.
    answer: c

28. The resting potential of a rod cell membrane in the dark is
    a. 0.
    b. more negative than a typical nerve cell membrane.
    c. about the same as a typical nerve cell membrane.
    d. less negative than a typical nerve cell membrane.
    answer: d

29. Unlike the light activation of visual rhodopsin, light activation of bacterial rhodopsin results in
    a. conversion of all-trans-retinal to 11-cis-retinal.
    b. conversion of all-trans-retinal to 13-cis-retinal.
    c. conversion of 11-cis-retinal to 13-cis-retinal.
    d. conversion of 9-cis-retinal to all-trans-retinal.
    answer: b.

30. Bacteriorhodopsin uses light energy to pump
    a. Na^+ ions out of the cell and H^+ ions into the cell.
    b. Na^+ ions into the cell and Cl^- ions out of the cell.
    c. H^+ ions out of the cell.
    d. Na^+ ions out of the cell and K^+ ions into the cell.
    answer: c

Chapter 30 Structures of Nucleic Acids and Nucleoproteins

1. In Frederick Griffith's classic 1928 experiment, smooth (S) strains of pneumococcus bacteria caused disease, whereas rough (R) strains were nonpathogenic. Griffith injected mixtures of these strains into susceptible mice. Which mice died?
   a. Those injected with heat killed S bacteria.
   b. Those injected with heat killed R bacteria.
   c. Those injected with a mixture of live R bacteria plus heat killed S bacteria.
   d. Those injected with a mixture of heat killed R bacteria and live R bacteria
   answer: c

2. The technical term for altering the genetic composition of one cell strain through the introduction of DNA from another strain is
   a. transfection
   b. transversion
   c. transduction
   d. transformation
   answer: d

3. If T2 phage are prepared in the presence of radioactive ^{32}P and ^{35}S,
   a. the protein is labeled with ^{32}P and the DNA is labeled with ^{35}S.
   b. the protein is labeled with both ^{32}P and ^{35}S, whereas the DNA is primarily labeled with ^{35}S.
   c. the DNA is labeled with ^{32}P and the protein is labeled with ^{35}S.
   d. the DNA is labeled with both ^{32}P and ^{35}S, whereas the protein is primarily labeled with ^{35}S.
   answer: c

4. In Hershey and Chase's experiment with radioactively labeled T2 phage,
   a. both DNA and protein from the phage entered the host cells during infection.
   b. neither DNA or protein from the phage entered the host cells during infection.
   c. only phage protein and not DNA entered the host cells during infection.
   d. only phage DNA and not protein entered the host cells during infection.
   answer: d

5. Of the following, which has never been observed as a carrier of genetic information?

   a. single-stranded DNA
   b. double-stranded DNA
   c. triplex DNA
   d. single-stranded RNA
   answer: c

6. The amount of DNA in a single diploid human cell is **about**
   a. 50,000-100,000 base pairs.
   b. 6×10^7 base pairs.
   c. 6×10^9 base pairs.
   d. 6×10^{11} base pairs.
   answer: c

7. Nucleotides are covalently linked together in a DNA strand through a bond between between
   a. the oxygen atom linted to the the 3'-carbon of one nucleotide and the 5' phosphate group of the next nucleotide in the chain.
   b. the 3' phosphate group of one nucleotide and the 5' phosphate group of the next nucleotide in the chain.
   c. a phosphorus atom linked to the 3' carbon of one nucleotide and the 5' carbon atom of the deoxyribose of the next nucleotide in the chain.
   d. the 3'-OH group of one nucleotide and the 3' phosphate group of the next nucleotide in the chain.
   answer: a

8. Chargaff's rules state that, for any double-stranded DNA,
   a. the amount of A + T is equal to the amount of G + C.
   b. the amount of A + G is equal to the amount of C + T.
   c. the amount of G is equal to the amount of A and the amount of T is equal to the amount of C.
   d. averaged over an entire organism, there are roughly equal amounts of A, T, G and C.
   answer: b

9. Which of the following is a true statement?
   a. A and T are paired by two hydrogen bonds linking the bases in a parallel orientation.
   b. G and C are paired by two hydrogen bonds linking the bases in an antiparallel orientation.
   c. G and C are paired by three hydrogen bonds linking the bases in an antiparallel orientation.
   d. G and C are paired by three hydrogen bonds linking the bases in a parallel orientation.
   answer: c

10. The length of a typical bacterial gene is 1000 base pairs. If this DNA were stretched into a single linear segment, how long would it be?
    a. 34 Å
    b. 1000 Å
    c. 3400 Å
    d. 10,000 Å
    answer: c

11. None of the following are typical base pairs found in DNA. However, if each of these were paired, which pair would have the greatest width?
    a. GA
    b. AC
    c. TG
    d. TC
    answer: a

12. The overall charge of a DNA molecule is
    a. positive.
    b. negative.
    c. uncharged.
    d. neutral because of equal numbers of positive and negative charges.
    answer: b

13. The DNA double helix structure is stabilized
    a. primarily by covalent bonds between adjacent chains.
    b. primarily by stacking interactions between bases.
    c. primarily by hydrogen bonds between opposite chains.
    d. by approximately equal contributions from stacking forces between adjacent bases and hydrogen bonds between opposite chains.
    answer: d

14. The sugar moieties in B DNA are
    a. planar.
    b. parallel to their respective bases.
    c. puckered in a C 2′ endo conformation.
    d. in a syn conformation with respect to their rotation about the purine C1-N9 glycosidic bonds.
    answer: c

15. Which of the following statements comparing the A and B forms of DNA is TRUE?
    a. Both are right handed helices.
    b. The major and minor groves are identical in both forms of DNA.
    c. The sugar puckers are identical in both forms of DNA.
    d. The width of the helix and the length of the helix per turn are identical in both forms of DNA.
    answer: a

16. Quadruplex (four stranded) DNA may exist in nature as a structural feature of
    a. nucleosomes.
    b. telomeres.
    c. centromeres.
    d. poly A regions of RNA.
    answer: b

17. The linking number (L) for a 240 base pair linear segment of B DNA with free ends in solution is
    a. 0
    b. -24
    c. 24
    d. 240
    answer: c

18. The enzyme responsible for introducing negative supercoils in bacterial DNA is
    a. DNA ligase.
    b. DNA gyrase.
    c. novobiocin.
    d. DNase.
    answer: b

19. The temperature at which double stranded DNA is denatured (the "melting temperature") is lowered
    a. with increasing salt concentration.
    b. with an increased percentage of GC base pairs.
    c. at either very high or very low pH.
    d. in the presence of histones.
    answer: c

20. C_ot curves are generated from samples of single stranded DNA which have been sheared to a uniform length of about 400 nucleotides and then allowed to reanneal. Which of the following samples would you expect to have the largest c_ot value?
    a. human satellite DNA
    b. poly AT
    c. viral DNA
    d. *E. coli* DNA
    answer: b

21. The current estimate for the number of genes in the human genome is
    a. 3000 genes.
    b. 50,000-100,000 genes.
    c. 1,000,000 genes.
    d. 3×10^9 genes.
    answer: b

22. The percentage of human DNA that actually codes for genes is
    a. $< 10\%$.
    b. about 25%.
    c. about 50%.
    d. about 90%.
    answer: a

23. If the base sequence of one strand of DNA is: 5′ AGACTCC 3′ , what is the sequence of its complementary strand?
    a. 5′ GGAGTCT 3′
    b. 5′ TCTGAGG 3′
    c. 5′ CCTCAGA 3′
    d. 5′ GGAGUCU 3′
    answer: a

24. The structure formed when DNA is wrapped around a bundle of 8 histone proteins is called a
    a. histosome.
    b. nucleosome.
    c. ribosome.
    d. nucleolus.
    answer: b

25. Histones are particularly rich in which of the following amino acids?
    a. serine and threonine
    b. glutamate and aspartate
    c. lysine and arginine
    d. valine and tryptophan
    answer: c

26. In contrast to eukaryotes, one of the components that stabilizes bacterial chromatin structure appears to be
    a. histones.
    b. RNA.
    c. Mg^{2+}.
    d. non-histone proteins.
    answer: b

27. The A form of DNA is preferred in
    a. prokaryotes.
    b. telomeres.
    c. RNA-DNA heteroduplexes.
    d. nucleosomes.
    answer: c

Chapter 31 DNA Replication, Repair and Recombination

1. Matthew Meselson and Frank Stahl used ^{15}N-labeled DNA and ^{14}N-labeled DNA to demonstrate that
   a. the two strands in DNA are antiparallel.
   b. DNA replication is semi-discontinuous.
   c. DNA replication is semi-conservative.
   d. DNA replication is bi-directional.
   answer: c

2. Meselson and Stahl started with *E. coli* cells containing pure heavy DNA (^{15}N-^{15}N-labeled DNA). After two generations of growth in ^{14}N medium, the cells contained
   a. 50% ^{15}N-^{15}N-labeled DNA and 50% ^{14}N-^{14}N-labeled DNA.
   b. 25% ^{15}N-^{15}N-labeled DNA; 50% ^{15}N-^{14}N-labeled DNA and 25% ^{14}N-^{14}N-labeled DNA.
   c. 50% ^{15}N-^{14}N-labeled DNA and 50% ^{14}N-^{14}N-labeled DNA.
   d. 25% ^{15}N-^{14}N-labeled DNA and 75% ^{14}N-^{14}N-labeled DNA.
   answer: c

3. Imagine that, instead of their actual results, Meselson and Stahl observed the following results: The cells were initially completely labeled with ^{15}N-DNA. After one generation of growth on ^{14}N-medium, they observed one band of double-stranded DNA with intermediate density. They heated their DNA to separate the strands and when the single strands were centrifuged in a CsCl gradient, they also yielded a single band of DNA with intermediate density. If they had gotten these results, Meselson and Stahl would have concluded that DNA replication is
   a. conservative.
   b. dispersive.
   c. semi-conservative.
   d. discontinuous.
   answer: b

4. In *E. coli*, each round of DNA replication starts at
   a. dnaA.
   b. oriC.
   c. a promoter site.
   d. the centromere.
   answer: b

5. Scientists believe that replication of the *E. coli* chromosome is **bidirectional on both strands** based on the following evidence:
   a. Autoradiography of ^{3}H labeled chromosomes show two θ (theta) structures during replication.
   b. About half of the replicated DNA is made in long strands while the other half is made in small pieces.
   c. After two rounds of replication in a medium that contains ^{3}H-labeled thymine, the θ (theta) structures are twice as strongly labeled as the rest of the chromosome.
   d. If growing bacteria are briefly exposed to ^{3}H-labeled thymine, and their chromosomes are examined by autoradiography, both forks in a θ (theta) structure are each intensely labeled.

   answer: d

6. Which of the following statements about DNA replication in *E. coli* is NOT CORRECT?
   a. If growing cells are exposed to radioactively labeled nucleotides for 2-10 seconds and then examined to determine the sizes of the growing DNA strands, about 50% of the strands are long stretches of DNA whereas the other 50% are shorter stretches, about 1000-2000 base pairs in length.
   b. If growing cells are exposed to radioactively labeled nucleotides for 1-2 minutes and then examined to determine the sizes of the growing strands, almost all of the DNA is in the form of long stretches.
   c. Synthesis of the leading strand proceeds in the 5′ to 3′ direction, whereas synthesis of the lagging strand is 3′ to 5′.
   d. Synthesis of the leading strand can occur continuously in the direction of the unwinding replication fork, whereas synthesis of the lagging strand occurs in discontinuous spurts in the opposite direction from the direction of the unwinding fork.

   answer: c

7. Okazaki fragments are
   a. short RNA oligonucleotides that act as starting points for DNA synthesis.
   b. short DNA oligonucleotides that are synthesized as part of the lagging strand.
   c. DNA oligonucleotides that are synthesized as part of the leading strand.
   d. fragments at the ends of chromosomes that have the sequence, GGTTG.

   answer: b

8. In the process of DNA replication which of the following enzymes acts first?
   a. helicase
   b. primase
   c. DNA polymerase
   d. DNA ligase

   answer: a

9. DNA polymerase requires all of the following EXCEPT:
   a. a DNA template.
   b. an RNA primer.
   c. Mg^{2+}.
   d. a growing chain with a free 5' phosphate end.
   answer: d

10. DNA pol I is the enzyme that was originally isolated from *E. coli* by Klenow in 1957. This enzyme has all of the following activities EXCEPT:
    a. 3' to 5' synthesis.
    b. 5' to 3' synthesis.
    c. 3' to 5' exonuclease.
    d. 5' to 3' exonuclease.
    answer: a

11. Which of the following proteins is NOT required at a replication fork?
    a. single strand DNA binding protein (SSB)
    b. gyrase
    c. primase
    d. helicase
    answer: b

12. In *E. coli*, after primer removal, the gap at the 5' end of the newly-synthesized DNA chains is filled by
    a. the 5' to 3' exonuclease activity of DNA pol I.
    b. the 5' to 3' exonuclease activity of DNA pol II.
    c. the 5' to 3' exonuclease activity of DNA pol III.
    d. the 3' to 5' exonuclease activity of DNA pol I.
    answer: a

13. In *E. coli*, it was originally believed that DNA pol I was the major polymerase responsible for synthesis of new DNA strands at each of the replication forks. However, subsequently it was shown that DNA polymerase III serves this function. The evidence that it is true includes all of the following EXCEPT:
    a. Mutant strains of *E. coli* have been isolated that have normal DNA replication, yet have only 1% of wild type DNA pol 1 activity.
    b. The turnover number (the number of nucleotides polymerized per minute per molecule) is 9000 for DNA pol III compared to 600 for DNA pol I and 30 for DNA pol II.
    c. Temperature sensitive DNA pol III mutants grow at low (permissive) temperatures, but not at high (non-permissive) temperatures.
    d. DNA pol III has 3' to 5' exonuclease activity but not 5' to 3' exonuclease activity.
    answer: d

14. Nick sealing by DNA ligase requires the formation of an enzyme-AMP intermediate. In *E. coli*, the AMP is supplied by
    a. NADH
    b. NAD⁺
    c. ATP
    d. cAMP
    answer: b

15. The chemistry of DNA synthesis involves
    a. a nucleophilic attack by the 3′ OH group of the entering dNTP molecule on the free 5′ phosphate group of the growing DNA chain.
    b. a nucleophilic attack by the 5′ OH group of the entering dNTP molecule on the free 3′ phosphate group of the growing DNA chain.
    c. a nucleophilic attack by the 3′ OH group of the growing chain on the alpha P atom of the free 5′ phosphate group of the entering dNTP molecule.
    d. a nucleophilic attack by the 3′ OH group of the growing chain on the gamma P atom of the free 5′ phosphate group of the entering dNTP molecule.
    answer: c

16. The sequence of the telomerase RNA in the ciliated protozoa tetrahymena is 5′ CAACCCAA 3′. Thus, this sequence codes for a repeating telomere sequence of
    a. 3′ GTTGGG 5′
    b. 5′ GTTGGG 3′
    c. 5′ TTGGGG 3′
    d. 5′ UUGGGG 3′
    answer: c

17. The SOS DNA repair system
    a. specifically repairs pyrimidine dimers.
    b. acts at GATC sites that are hemi-methylated.
    c. is normally repressed by lexA.
    d. requires visible light.
    answer: c

18. UV light damages DNA by
    a. inducing cross links between strands.
    b. repressing DNA repair systems.
    c. stimulating nucleotide deamination.
    d. stimulating dimerization of adjacent pyrimidine bases.
    answer: d

19. A mutation that leads to defective DNA polymerase III 3′ to 5′ exonuclease activity
    a. would be lethal.
    b. would result in the inability to excise RNA primers.
    c. would result in a much higher than normal mutation rate.
    d. would have only a slight deleterious effect on DNA replication.
    answer: c

20. Which of the following enzymes or viruses does not have reverse transcriptase activity?
    a. telemerase.
    b. eukaryotic RNase H.
    c. primase.
    d. HIV.
    answer: c

21. Recombination of homologous double-stranded DNA requires
    a. nicks in each of two complementary strands.
    b. nicks in all fours strands.
    c. a single nick in one of the strands.
    d. nicks in each of two homologous single strands.
    answer: d

Question 22 refers to the following diagram.

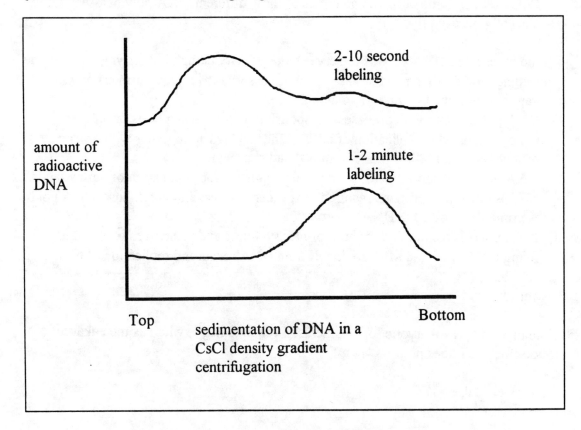

195

22. The experiment illustrated above demonstrates that
   a. Okazaki fragments are joined within 1-2 minutes.
   b. RNA primers are removed within 1-2 minutes.
   c. the time it takes for one round of chromosome replication in *E. coli* is about 1-2 minutes.
   d. the time it takes to incorporate radioactivity into DNA is 1-2 minutes.
   answer: a

23. Replication origins in yeast are called
   a. oriC sequences.
   b. centromeres.
   c. T antigen binding sites.
   d. ARS sequences.
   answer: d

24. Eukaryotic DNA pol α has primase activity, 5′ to 3′ polymerase activity and 3′ to 5′ exonuclease activity, but low processivity.

   • DNA pol δ has no primase activity. It does, however, have 5′ to 3′ polymerase activity and 3′ to 5′ exonuclease activity and it has high processivity.
   • A protein called MF1 has 5′ to 3′ exonuclease activity.
   • Consistent with the observed activities of these different DNA polymerases, it is currently believed that

   a. the function of DNA pol δ is to serve as the primary elongation enzyme on both the leading strand and lagging strand. DNa pol α serves to remove primers and fill gaps between the Okazaki fragments.
   b. DNA pol α, along with primase, is responsibl   e for the synthesis of primers, DNA pol δ is involved in replication of both strands and MF1 (together with other proteins) is required for primer removal and gap filling.
   c. DNA pol α is responsible for both primer synthesis and replication of both strands. MF1 is one of the proteins required for primer removal and gap filling. DNA pol δ is primarily involved in DNA repair.
   d. DNA pol α is responsible for both primer synthesis and primer removal and gap filling DNA pol δ and MF1 are involved in DNA repair and replication of both strands.
   answer: b

25. True or False? Prokaryotic topoisomerase I requires Mg^{2+}, whereas the eukaryotic topoisomerase I does not.

   answer: false

26. True or False? Type II topoisomerase enzymes from prokaryotes produce transient breaks in both of the strands of a DNA double helix, whereas Type II topoisomerases in eukaryotes produce a transient single strand break.

answer: true

27. In *E. coli*, the protein that binds to the origin of replication to initiate each round of DNA replication is called
   a. DNA A.
   b. DNA B.
   c. DNA C.
   d. primase.
   answer: a

28. The β subunit of *E. coli* DNA polymerase III dramatically increases the processivity of the enzyme. It is believed to act by a mechanism called
   a. a rolling circle.
   b. D-loop formation.
   c. a migrating enzyme mechanism.
   d. a sliding clamp.
   answer: d

29. Mutations in recA completely eliminate recombination between homologous chromosomes. Which of the following statements is NOT TRUE about the functions of the recA protein?
   a. RecA catalyzes the formation of D loops in the presence of single-stranded DNA and negatively-supercoiled circular DNA.
   b. RecA can catalyze the formation of strand exchange between two double-stranded helices, provided at least one of the helices is gapped on one strand.
   c. RecA catalyzes the formation of the initial incisions in DNA duplex structures that are required to initiate recombination.
   d. When bound to a single-stranded DNA fragment, recA can stimulate self cleavage of the lexA protein involved in repression of the SOS repair system.
   answer: c

30. Short Answer: Briefly explain how the mismatch repair system recognizes the difference between a parent strand and a newly-synthesized daughter strand.

   answer: The A's are methylated in all of the GATC sequences of the parent strand. However, it takes several minutes for these sequences to be methylated in the newly-synthesized strands. Thus, the mismatch repair enzymes recognize strands with methylated GATC sequences as parent strands and strands with unmethylated GATC sequences as newly-synthesized daughter strands.

Chapter 32 DNA Manipulation and its Applications

1. A reaction mixture for DNA sequencing using the Sanger method requires all of the following EXCEPT:
   a. dideoxy NTP's.
   b. deoxy NTP's.
   c. NTP's.
   d. DNA polymerase.
   answer: c

2. The DNA sequence that corresponds to the gel shown at the right is:
   a. 5′ CCTAGGG 3′
   b. 5′ GGATCCC 3′
   c. 5′ GGGATCC 3′
   d. 5′ CCCTAGG 3′
   answer: c

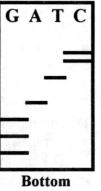

Bottom

3. Each of the fragments generated by a Sanger DNA sequencing procedure
   a. contains a single dideoxynucleotide incorporated at its 3′ end.
   b. contains a single dideoxynucleotide incorporated at its 5′ end.
   c. contains a single dideoxynucleotide incorporated randomly at the appropriate A,G,T or C nucleotide sites within its sequence.
   d. contains multiple dideoxynucleotides incorporated at the appropriate A, G, T or C nucleotide sites within its sequence.
   answer: a

4. In the Sanger DNA sequencing procedure, gel electrophoresis separates DNA fragments that have
   a. different ratios of purines/pyridinines.
   b. different charge to mass ratios.
   c. different sizes.
   d. different numbers of incorporated dideoxynucleotides.
   answer: c

5. In a biochemistry lab experiment, your group sets up a PCR amplification procedure. You start with template DNA in a test tube; add two primers and all of the deoxynucleotide triphosphates. What have you forgotten?
   a. dideoxynucleotide triphosphates.
   b. DNA polymerase.
   c. ATP.
   d. restriction enzymes.
   answer: b

6. The amount of DNA that can be produced from a single double-stranded template DNA molecule after only 10 rounds of cycling in a PCR procedure is
   a. 20 molecules.
   b. 100 molecules.
   c. 1024 molecules.
   d. about 1,000,000 molecules.
   answer: c

7. Which of the following sequences exhibits characteristics that are typical of the recognition sites for most known restriction enzymes?
   a. AAGCTT
   b. CCCTTT
   c. GCAT
   d. AAAAT
   answer: a

8. Cosmids have an advantage over plasmids and bacteriophage as DNA vectors because
   a. they are resistant to most restriction enzymes.
   b. they replicate faster than either plasmids or bacteriophage.
   c. they induce lysis of the host bacterial cells.
   d. they can be used to clone segments of DNA that are between 25 kb and 50 kb in length.
   answer: d

9. Which of the following vectors is used to clone the smallest DNA segments?
   a. bacteriophage λ.
   b. cosmids.
   c. YACs.
   d. plasmids.
   answer: d

10. A genomic library is
    a. a collection of recombinant host cells that each contain different fragments of the genome of a particular organism.
    b. a collection of DNA copies of all of the messenger RNA molecules made by a particular cell.
    c. a collection of all of the RNA transcripts made by a particular tissue or cell type.
    d. a computer file that contains the DNA sequence for the entire genome of a particular organism.
    answer: a

11. The size of the final product in a PCR reaction is determined by
    a. the length of DNA for which the ends are defined by the two primers.
    b. the length of the template chain.
    c. the amount of time allowed for each cycle of heating and cooling.
    d. the processivity of the DNA polymerase.
    answer: a

12. The ampicillin resistance gene in the pBR322 vector plasmid is useful because
    a. it enhances the viability of the cells that incorporate the pBR322 plasmid.
    b. it contains a unique restriction site. Thus, if any foreign gene is inserted in the plasmid, cells that contain the recombinant DNA will not be ampicillin resistant.
    c. it contains a unique restriction site. Thus, if any foreign gene is inserted in the plasmid, only cells that contain the recombinant DNA will be ampicillin resistant.
    d. It allows the pBR322 plasmid to be used as a shuttle vector that can replicate in either yeast or *E. coli*.
    answer: b

13. In a cloning procedure that utilizes the vector plasmid pBR322, each host cell typically produces
    a. 1 plasmid.
    b. about 100 copies of the vector plasmid.
    c. 1000-2000 copies of the vector plasmid.
    d. about 10^6 copies of the vector plasmid.
    answer: c

14. A restriction enzyme that recognizes a four base sequence gives an average fragment size of
    a. 256 base pairs.
    b. 1024 base pairs.
    c. 4096 base pairs.
    d. 16,384 base pairs.
    answer: a

15. In cosmids, the cos sites are useful because
    a. they act as an origin of replication.
    b. they have been modified to contain many unique restriction sites for foreign gene insertion.
    c. they allow the cosmids to replicate in yeast cells.
    d. they allow the cosmids to be packaged into λ phage heads for infection into host cells.
    answer: d

16. The 3′ transcriptional control regions of the human β globin genes could be found
    a. in a human genomic library.
    b. in a cDNA library of red blood cells.
    c. in a cDNA library of bone marrow cells.
    d. by PCR amplification of the mRNA in red blood cells.
    answer: a

17. To detect RFLP's, researchers use a technique called
    a. Eastern blotting.
    b. Northern blotting.
    c. Southern blotting.
    d. Western blotting.
    answer: c

18. RFLP's result from
    a. DNA sequence differences that affect restriction enzyme recognition sites.
    b. DNA sequence differences in transcriptional control regions of essential genes.
    c. Mutations that alter the length of mRNA transcripts.
    d. Mutations that affect the sensitivity of RNA fragments to specific ribonucleases.
    answer: a

19. Mutants that are defective in thymidine kinase do not grow on HAT medium. Gancyclovir inhibits the thymidine kinase coded by herpes-type viruses, but does not inhibit the thymidine kinase from mamalian cells. Thus, an appropriate strategy for creating and selecting mouse cells that incorporate the human β globin gene is to,

    a. 1) Start with tk⁻ cells. 2) Transform the cells using a vector plasmid that contains both the herpes tk gene and the human β globin gene. 3) Plate on HAT medium 4) Add gancyclovir.
    b. 1) Start with tk⁻ cells. 2) Transform the cells using a vector plasmid that contains both the herpes tk gene and the human β globin gene. 3) Add gancyclovir 4) Plate on HAT medium.
    c. 1) Start with tk⁺ cells. 2) Transform the cells using a vector plasmid that contains both the herpes tk gene and the human β globin gene. 3) Plate on HAT medium.
    d. 1) Start with tk⁻ cells. 2) Transform the cells using a vector plasmid that contains both the herpes tk gene and the human β globin gene. 3) Plate on HAT medium.
    answer: d

20. Teratocarcinoma cells are cancerous cells that can revert back to normal growth
    a. when placed in a glass capillary tube.
    b. when injected subcutaneously into an appropriate host.
    c. when injected into a blastocyst.
    d. when transformed with a tk gene from a Herpes simplex virus.
    answer: c

21. The selection step in Southern blotting involves hybridization of
    a. a single-stranded ^{32}P DNA probe with DNA bands on an agarose gel.
    b. a single-stranded ^{32}P DNA probe with DNA bands on a cellulose nitrate sheet.
    c. a double-stranded ^{32}P DNA probe with DNA bands on an agarose gel.
    d. a double-stranded ^{32}P DNA probe with DNA bands on a cellulose nitrate sheet.
    answer: b

22. Researchers started with cDNA for the human β globin gene and used that, together with a genomic library, to construct a series of probes to explore the regions flanking the β globin gene. This procedure is called
    a. RFLP analysis.
    b. chromosome jumping.
    c. chromosome walking.
    d. DNA footprinting.
    answer: c

Questions 23 and 24 refer to the experiment illustrated in the diagram shown below.

Defective gene segment A	neoR gene	Defective gene segment B	Herpes tk gene

23. In an experiment designed to select for targeted recombination of a defective cloned gene fragment with the homologous location in a mammalian chromosome, a cloning vector was constructed by splitting the cloned gene and inserting genes for neomycin resistance (neoR) and thymidine kinase form Herpes Simplex (tk) as shown in the diagram above. In this experiment, cells that undergo homologous recombination
    a. carry the defective version of the gene and are sensitive to neomycin.
    b. carry the defective version of the gene and are resistant to neomycin.
    c. carry the intact wild type gene and are sensitive to neomycin.
    d. carry the intact wild type gene and are resistant to neomycin.
    answer: b

24. The host cells in this experiment already contained a functional mammalian tk gene. The purpose of the Herpes tk gene in this experiment was to
    a. allow only cells that had undergone homologous recombination to survive in HAT medium.
    b. cause cells that had undergone homologous recombination to be susceptible to gancyclovir.
    c. cause cells that had either undergone homologous recombination of had not undergone any recombination to be susceptible to gancyclovir.
    d. cause cells that had undergone random recombination, but not cells that had undergone homologous recombination to be susceptible to gancyclovir.
    answer: d

202

25. Which of the following types of cloning vectors would you prefer if you needed to create a genomic library of the entire human genome?
   a. plasmids.
   b. bacteriophage λ clones.
   c. cosmids.
   d. YACs.
   answer: d

26. A jumping library is usually used in conjunction with a complementary
   a. cDNA library.
   b. linking library.
   c. cosmid library.
   d. T1 library.
   answer: b

27. Which of the following methods would you start with if you were trying to isolate a gene that corresponds to a protein that is highly expressed only in lung tissue?
   a. Use PCR to amplify the gene.
   b. Isolate a clone from a cDNA library of lung tissue.
   c. Use a chromosome jumping procedure.
   d. Isolate the gene from a human genomic library.
   answer: b

28. Which of the following enzymes is needed to produce cDNA?
   a. RNA polymerase.
   b. a restriction endonuclease.
   c. DNA polymerase I.
   d. reverse transcriptase.
   answer: d

29. In an experiment designed to demonstrate the feasibility of gene therapy for cancer, cells called tumor infiltrating lymphocytes (TILs) were isolated from a cancer patient. (TILs are T cells that attack tumor cells.) These cells were transfected using a retrovirus vector that contained a neomycin resistance gene. The transfected cells were returned to the patient. After several weeks, the researchers wanted to see if the neomycin resistance gene could still be detected in this patient. If you were doing this experiment, which of the following methods would you use to detect the recombinant neomycin resistance gene in the TIL cells?
   a. Perform a chromosome walking procedure on the DNA from the TIL cells using a cDNA probe.
   b. Clone the DNA from the TIL cells into a retrovirus vector.
   c. Use PCR and two primers specific to the neomycin resistance gene to amplify a portion of the DNA from the TIL cells.
   d. Clone the DNA from the TIL cells into a YAC vector.
   answer: c

30. The protein product of the gene that is defective in cystic fibrosis is
   a. a hormone receptor protein.
   b. a tumor supressor protein.
   c. a Na^+, K^+-ATPase.
   d. a membrane protein that functions as a channel for halides, water and other small solutes.
   answer: d

Chapter 33 RNA Synthesis and Processing

1. The average half-life of a mRNA molecule in a bacterial cell is about
   a. 1-3 seconds.
   b. 1-3 minutes.
   c. 1-3 hours.
   d. 10 hours.
   answer: b

2. A typical tRNA molecule is
   a. 75-90 nucleotides.
   b. about 500 nucleotides.
   c. about 1500 nucleotides.
   d. about 2000-6000 nucleotides.
   answer: a

3. The 30S ribosomal subunit contains
   a. 5S RNA and 16S RNA.
   b. 5S RNA and 23S RNA.
   c. a single 16S RNA molecule.
   d. 30S RNA.
   answer: c

4. The 3′ OH ends of all tRNA molecules have the sequence
   a. AAA
   b. AAUAA
   c. ACC
   d. CCA
   answer: d

5. A GC-rich stem sloop followed by six U's is indicative of
   a. a heat shock promoter sequence.
   b. a highly methylated sequence.
   c. a 5S rRNA sequence.
   d. a terminator sequence.
   answer: d

6. RNA polymerase binds at specific sites on DNA called
   a. Shine-Dalgarno sequences.
   b. promoters.
   c. rho factors.
   d. initiators.
   answer: b

7. Without its (σ) sigma subunit, *E. coli* RNA polymerase
   a. cannot elongate an RNA chain.
   b. cannot respond to rho-dependent termination factors.
   c. cannot initiate transcription unless the DNA has single-stranded nicks.
   d. can polymerize RNA but, cannot use DNA as a template for RNA synthesis.
   answer: c

8. Which of the following statements about rho-dependent termination is NOT CORRECT?
   a. Rho moves along RNA in a 5′ to 3′ direction.
   b. Rho movement along RNA requires ATP.
   c. The helicase activity of the rho protein is not essential for its function.
   d. Rho causes termination only at pause or termination sites.
   answer: c

9. At low concentrations, α amanitin inhibits
   a. all RNA polymerases.
   b. eukaryotic RNA polymerase I.
   c. eukaryotic RNA polymerase II.
   d. all prokaryotic RNA polymerases.
   answer: c

10. In eukaryotes, the RNA polymerase II binding site
    a. is a consensus sequence located at -10 nucleotides upstream from the transcription start site.
    b. is a consensus sequence usually located at -20 to -30 nucleotides upstream from the transcription start site.
    c. has two consensus sequences: one at -35 and one at -10 nucleotides upstream form the transcription start site.
    d. is located at the UAS site about 40-200 nucleotides upstream from the transcription start site.
    answer: b

12. All of the following statements about the binding of the TATA binding protein (TBP) to the "TATA" box are true EXCEPT
    a. It causes the DNA to bend by about 80°.
    b. It causes the DNA double helix to unwind by about 110°.
    c. Along with other factors, it binds in a complex with RNA polymerase II.
    d. It phosphorylates the CTD region of RNA polymer.
    answer: d

13. The first step in the formation of the RNA polymerase II initiation complex is
    a. the binding of (transcription factor) TFIIB to the TATA box.
    b. the binding of RNA polymerase II to the TATA box.
    c. the binding of (transcription factor) TFIID to the TATA box.
    d. bending of the DNA through an angle of 80° at the TATA element.
    answer: c

14. In eukaryotes, mRNA is transcribed by
    a. RNA polymerase I.
    b. RNA polymerase II.
    c. RNA polymerase III.
    d. both RNA polymerase I and II.
    answer: b

15. In eukaryotes, the 5S RNA genes are transcribed by
    a. RNA polymerase I.
    b. RNA polymerase II.
    c. RNA polymerase III.
    d. both RNA polymerase I and II.
    answer: c

16. Rifampacin inhibits
    a. prokaryotic RNA polymerase.
    b. RNA polymerase I.
    c. RNA polymerase II.
    d. RNA polymerase III.
    answer: a

17. Both prokaryotic and eukaryotic RNA polymerases are inhibited by
    a. actinomycin D.
    b. low concentrations ethydium bromide.
    c. α aminitin.
    d. rifampacin.
    answer: a

18. The 5' end of an eukaryotic mRNA molecule is
    a. CCA.
    b. a poly A sequence.
    c. 7-methyl guanosine.
    d. AAUAA.
    answer: c

19. Which of the following statements is a CORRECT overview of mRNA splicing?
    a. During splicing, the 5′ cap is removed, the introns are removed, the poly A tail is removed and the exons are joined together.
    b. During splicing, the 5′ cap and the poly A tail are retained, the introns are removed and the exons are joined together.
    c. During splicing, the 5′ cap is removed, the exons are removed, the poly A tail is removed and the introns are joined together.
    d. During splicing, the 5′ cap and the poly A tail are retained, the exons are removed and the introns are joined together.
    answer: b

20. Which of the following statements accurately describes the sequence of events during splicing?
    a. During splicing, U1 binds to the 5′ splice site then U2 binds to the branch site. The 2′ OH group of an adenosine attacks the 5′ phosphate of a guanine residue; a lariat is formed and the exons are ligated.
    b. During splicing, U1 binds to the 3′ splice site then U2 binds to the branch site. The 5′ OH group of an adenosine attacks the 2′ phosphate of a guanine residue; a lariat is formed and the exons are ligated.
    c. During splicing, U1 binds to the 5′ splice site then U2 binds to the branch site. The 5′ OH group of an adenosine attacks the 3′ phosphate of a guanine residue; a lariat is formed and the exons are ligated.
    d. During splicing, U1/U2 binds to the 5′ splice site then U5/U4/U6 binds to the branch site. The 2′ OH group of an adenosine attacks the 5′ phosphate of a guanine residue; a lariat is formed and the exons are ligated.
    answer: a

21. The catalytically active component of ribonuclease P is
    a. U1 snRNP.
    b. RNA.
    c. GTP.
    d. nucleolin.
    answer: b

22. RNA editing to insert nucleotides at specific locations in an mRNA transcript requires
    a. 7S RNA.
    b. spliceosomes.
    c. RNase H.
    d. guide RNA.
    answer: d

6. Mitochondria have a genetic code that is slightly different from the standard genetic code. Only 24 different tRNA molecules are utilized by mitochondria compared to the 32 tRNAs that are <u>minimally</u> required to translate all 61 codons. The explanation for this difference is that
   a. mitochondria do not use all 61 codons.
   b. the wobble rules are different in mitochondria compared to nuclear codons.
   c. mitochondria typically make only 10-20 different proteins and these proteins completely lack several of the 20 amino acids.
   d. tRNA molecules in mitochondria do not contain modified bases.
   answer: b

7. The modified base, inosine, is derived from A by deamination. When inosine is in the 5′ position of an anticodon it can basepair with
   a. A or G
   b. G or U
   c. G or C
   d. U, C or A
   answer: d

8. In the first step of the synthesis of an aminoacyl tRNA molecule,
   a. the carboxyl group from the amino acid attaches to the phosphate group of AMP to form an aminoacyl-AMP intermediate.
   b. the carboxyl group from the amino acid attached to the phosphate group of the 5′ end of a tRNA molecule to form an aminoacyl tRNA intermediate.
   c. the carboxyl group from the amino acid attaches to the α phosphate of GTP to form an aminoacyl-GMP intermediate.
   d. the carboxyl group from the amino acid attaches to the α phosphate of ATP to form an aminoacyl-AMP intermediate.
   answer: d

9. The overall error frequency of aminoacylation is
   a. about 10%.
   b. about 1%.
   c. about 1 in 10,000.
   d. about 1 in 10^9.
   answer: c

10. From the crystal structure of the complex between tyrosine and its corresponding aminoacyl synthase (tyrosyl synthase), it appears that enzyme recognition of the appropriate amino acid
    a. is determined primarily by the formation of specific hydrogen bonds between the enzyme and the amino acid.
    b. requires simultaneous binding of the amino acid and the tRNA.
    c. is determined primarily by hydrophobic interactions between the tRNA molecule and the amino acid.
    d. is determined by a second "proofreading" enzyme.
    answer: a

11. In their active form during translation, amino acids are attached to their cognate tRNA molecules through
    a. an ester linkage to the 5′ phosphate of a terminal C residue of the tRNA molecule.
    b. an ester linkage to the 3′ hydroxyl group of a terminal C residue of the tRNA molecule.
    c. an ester linkage to the 2′ hydroxyl group of a terminal C residue of the tRNA molecule.
    d. an ester linkage to the 3′ hydroxyl group of a terminal A residue of the tRNA molecule.
    answer: d

12. In prokaryotes, the ribosome recognizes and binds to the mRNA at a sequence called the Shine-Dalgarno sequence which is complementary to
    a. a sequence that is part of the tRNA^{Met} molecule.
    b. the 16S rRNA.
    c. the 23S rRNA.
    d. the initiation codon.
    answer: b

13. The catalyst for peptide bond formation during protein synthesis is
    a. EF-Tu.
    b. EF-Ts.
    c. EF-G.
    d. part of the ribosome.
    answer: d

14. Of the following, which binds to a free aminoacyl tRNA molecule?

    a. EF-Ts•EF-Tu•GDP.
    b. EF-Ts•EF-Tu.
    c. EF-Tu•GTP.
    d. EF-Ts•EF-Tu•GTP.
    answer: c

15. During elongation, it is thought that deacylated tRNA binds to a site on the ribosome called
    a. the A site.
    b. the E site.
    c. the P site.
    d. the Shine-Dalgarno sequence.
    answer: b

16. Which of the following factors does not bind directly to GTP?
    a. EF-Ts.
    b. EF-Tu.
    c. IF-2.
    d. IF-3.
    answer: a

17. The function of EF-G is
    a. to move the complex between a tRNA molecule and a growing peptide chain from the A site to the P site.
    b. to transfer a growing peptide chain from one tRNA molecule to another.
    c. to bring a new aminoacyl-tRNA molecule to its appropriate site in the ribosome.
    d. to displace EF-Ts from EF-Tu.
    answer: a

18. Dyptheria toxin specifically interacts with
    a. EF-Tu in prokaryotic cells.
    b. EF-2 in eukaryotic cells.
    c. EF-G in prokaryotic cells.
    d. EF-Ts in prokaryotic cells.
    answer: b

19. GMPPCP is a non-hydrolyzable analog of GTP that can substitute for GTP in binding to elongation proteins. Upon addition of GMPPCP to a complex of EF-Ts with EF-Tu,
    a. EF-Ts would be displaced, leaving an EF-Tu-GMPPCP complex.
    b. the complex would dissociate to form free EF-Ts, free EF-Tu and free GMPPCP.
    c. EF-Tu would be displaced, leaving an EF-Ts-GMPPCP complex.
    d. a stable complex would form between EF-Ts•EF-Tu and GMPPCP.
    answer: a

20. In *E. coli*, termination codons are recognized by
    a. termination tRNA molecules.
    b. one of two proteins called RF-1 and RF-2.
    c. the RF-3 protein.
    d. fMET-tRNA^{fMet}
    answer: b

21. Streptomycin specifically binds to
    a. the small ribosomal subunit in prokaryotes.
    b. the large ribosomal subunit in prokaryotes.
    c. the signal recognition particle (SRP).
    d. EF-G in prokaryotes.
    answer: a

22. In *E. coli*, the gene for the release factor RF-2 is not translated to make RF-2
    a. unless there is a frame shift that induces the ribosome to ignore a normal stop codon.
    b. unless there is a translational jumping event in which there is a skip of 50 bases downstream on the mRNA molecule.
    c. unless there is a chaperone protein to interact with the nacent RF-2.
    d. unless tetracycline is bound to the ribosome.
    answer: a

23. It is believed that, in order to be transported across the mitochondrial membrane, proteins must often be
    a. glycosylated.
    b. unfolded from their native conformation.
    c. targeted for transport by stop-transfer or signal-anchor sequences.
    d. bound to a specific mitochondrial signal recognition particle.
    answer: b

24. Transport across the endoplasmic reticulum membrane requires all of the following EXCEPT
    a. 7S RNA
    b. GTP
    c. the SRP (signal recognition particle) receptor
    d. PCB's (polypeptide chain binding proteins)
    answer: d

25. Lysosomal proteases are called
    a. cathepsins
    b. ubiquitins
    c. La proteases
    d. chaperones
    answer: a

214

26. In *E. coli*, the function of many heat shock proteins such as Hsp70 and TRIC is to
    a. bind to proteins and target them for denaturation.
    b. act as translation termination factors.
    c. act as chaperones to stabilize proper folding of cell proteins, both as they are translated and upon release from the ribosome.
    d. induce translation jumping and thus, alter the reading frame during translation.
    answer: c

27. Ubiquitin targets proteins for degradation by
    a. transporting them to lysosomes.
    b. phosphorylating them on specific serine, threonine or tyrosine residues.
    c. covalently binding to them.
    d. catalyzing proteolysis at specific arginine or lysine residues.
    answer: c

Chapter 35 Regulation of Gene Expression in Prokaryotes

1. The percentage of the approximately 3000 genes in E. coli that are actually being transcribed at any given time during active growth is
   a. < 1%
   b. about 5%
   c. about 30%
   d. about 50%
   answer: b

Questions 2-6 refer to the diagram shown below:

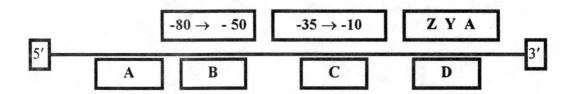

2. The lactose repressor protein binds to a location within region
   a. a
   b. b
   c. c
   d. d
   answer: c

3. RNA polymerase binds to region
   a. a
   b. b
   c. c
   d. d
   answer: c

4. The repressor (i) gene is located within region
   a. a
   b. b
   c. c
   d. d
   answer: a

5. The primary operator location is within region
   a. a
   b. b
   c. c
   d. d
   answer: c

6. The CAP binding location is within region
   a. a
   b. b
   c. c
   d. d
   answer: b

7. In *E. coli*, β galactosidase genes are not transcribed if there is adequate glucose to meet metabolic needs. This is an example of a general gene regulation mechanism called
   a. induction.
   b. feedback inhibition.
   c. the stringent response.
   d. catabolite repression.
   answer: d

8. In the presence of lactose plus glucose, *E. coli*, β galactosidase genes are not transcribed due to
   a. repressor binding to the operator region.
   b. cAMP-CAP binding to the operator region.
   c. lack of inducer.
   d. lack of cAMP which is needed to form a cAMP-CAP complex.
   answer: d

9. The o^c mutation results in a modification of the DNA sequence in the operator region of the lac operon. Repressor molecules cannot bind to this altered sequence. Thus, in this case,
   a. the mutant *E. coli* cell is able to utilize lactose even in the absence of glucose.
   b. the mutant *E. coli* cell is not able to utilize lactose even in the absence of glucose.
   c. the mutant *E. coli* cell transcribes the β galactosidase genes in the absence of both glucose and lactose.
   d. the presence of lactose prevents transcription of the β galactosidase genes even in the presence of glucose.
   answer: c

10. A virus or plasmid can be used to introduce a second copy of the lac operon into *E. coli* cells. If the second copy contains a mutation that prevents the synthesis of the repressor protein (an i⁻ mutation) and the original bacterial chromosome is wild type,
   a. β galactosidase transcription would be constitutive.
   b. there would be no change in the phenotype of the cell.
   c. the cell would loose the ability to respond to the inducer.
   d. the cell would loose the ability to exhibit catabolite repression.
   answer: b

11. The inducer of the lac operon is
   a. cAMP
   b. glucose
   c. CAP
   d. a metabolite of lactose
   answer: d

12. If the I gene (repressor gene) were moved downstream of the lactose operon,
   a. the structural genes would become inactivated.
   b. the operator region would become inactivated.
   c. the structural genes would be constitutively expressed.
   d. there would be no effect on the cell's ability to regulate the expression of the lactose operon.
   answer: d

13. The operon hypothesis was first proposed by
   a. Jacob and Monod.
   b. Gilbert and Muller-Hill.
   c. Lederman, DeVries and Zubay.
   d. Watson and Crick.
   answer: a

14. A group of biochemistry lab students constructs a synthetic operon in which they delete the structural genes for lactose utilization. They replace these genes with the structural genes for alanine biosynthesis, just downstream of all of the regulatory sequences for the lac operon. To induce expression of their recombinant alanine genes, they need to add _____ to their growth medium.
   a. alanine
   b. glucose
   c. allolactose
   d. β galactosidase
   answer: c

15. The trp repressor similar to the lac repressor in all of the following aspects EXCEPT
    a. both DNA binding sites have dyad symmetry.
    b. both of the repressor proteins bind to DNA through helix-turn-helix motifs.
    c. both of the repressor proteins are synthesized constitutively.
    d. each repressor binds to DNA only in the presence of an inducer.
    answer: d

16. In the regulation of the expression of *E. coli* genes for tryptophan biosynthesis, attenuation of the trp operon occurs as a result of
    a. an RNA hairpin loop that occurs within the leader sequence of the trp transcript only under typtophan-starved conditions.
    b. an RNA hairpin loop that occurs within the leader sequence of the trp transcript only in the presence of excess typtophan.
    c. binding of tryptophan to the trp repressor protein.
    d. the stringent response.
    answer: b

17. In *E. coli*, under conditions of general amino acid deprivation, rRNA synthesis ceases abruptly. This response is called
    a. attenuation.
    b. the stringent response.
    c. general catabolite repression.
    d. anti-attenuation.
    answer: b

18. Binding of ppGpp to *E. coli* RNA polymerase
    a. increases the affinity of the RNA polymerase for the promoters of rRNA, tRNA and ribosomal protein genes.
    b. increases the affinity of the RNA polymerase for the promoters of all *E. coli* genes.
    c. decreases the affinity of the RNA polymerase for the promoters of rRNA, tRNA and ribosomal protein genes.
    d. decreases the affinity of the RNA polymerase for the promoters of all *E. coli* genes.
    answer: c

19. The spoT gene controls the rate of breakdown of ppGpp thus,
    a. spoT⁻ mutants continue to synthesize high levels of rRNA under amino acid starvation conditions.
    b. In spoT⁻ mutants, amino acid starvation leads to reduction of rRNA synthesis, however the rate of rRNA synthesis increases only very slowly upon readdition of amino acids.
    c. spoT⁻ mutants synthesize low levels of rRNA even in the presence of excess amino acids.
    d. spoT⁻ mutants synthesize low levels of rRNA in the presence of excell amino acids but, high levels of rRNA under amino acid starvation conditions.
    answer: b

20. In *E. coli* under rapid growth conditions, all of the following factors contribute to high levels of rRNA synthesis EXCEPT:
   a. There are two promoters for the rRNA genes.
   b. There are 7 copies of the rRNA genes.
   c. The leader region of the rRNA transcipts contains several sequences that have the potential to form stem-and-loop structures that increase the levels of rRNA synthesis.
   d. There is an AT-rich region called the UP element upstream of the promoter site that interacts wth RNA polymerase to increase the levels of rRNA synthesis.
   answer: c

21. Which of the following is NOT a function of the cI repressor in bacteriophage λ?
   a. At low concentrations it binds to RNA polymerase at the nut_L site and acts as an antiterminator.
   b. At moderate concentrations it promotes its own continued expression.
   c. At high concentrations it inhibits overproduction of itself.
   d. It represses expression of genes involved in the lytic cycle.
   answer: a

22. UV light causes the bacteriophage λ prophage to enter the lytic cycle by
   a. damaging the viral DNA.
   b. activating the phage N protein.
   c. damaging the phage N protein.
   d. activating the host recA protease which, in turn, cleaves the cI repressor.
   answer: d

23. The bacteriophage λ protein, cro, binds to exactly the same region on DNA as
   a. N
   b. Q
   c. cI
   d. cII
   answer: c

24. Of the following bacteriophage λ genes, which is activated first?
   a. N
   b. cI
   c. Q
   d. int
   answer: a

25. The N protein from bacteriophage λ binds to
    a. The nut_L and nut_R sites.
    b. P_L and P_R.
    c. RNA polymerase.
    d. t_L and t_{R1}.
    answer: c

26. Proteins that bind to DNA through helix-turn-helix motifs typically have two binding sites that are
    a. separated by about 10 nucleotides on opposite sides of the DNA helix.
    b. separated by at least 100 nucleotides.
    c. are on the same sides of the DNA helix, bound to adjacent minor grooves.
    d. are on the same sides of the DNA helix, bound to adjacent major grooves.
    answer: d

27. Most helix-turn-helix proteins share all of the following features EXCEPT:
    a. dyad symmetry.
    b. the requirement for an essential zinc ion.
    c. frequent participation of glutamine or asparagine side chains in making contacts with the DNA base pairs.
    d. frequent use of glycine in the turn between the two helices.
    answer: b

28. The reason for the presence of adjacent inverted repeat sequences in the DNA binding sites of most helix-turn-helix proteins is:
    a. The DNA must form a hairpin loop in order for the helix-turn-helix protein to bind.
    b. Most of these proteins contain two essentially identical helix-turn-helix motifs which bind to adjacent "mirror image" DNA binding sites.
    c. RNA transcribed from these locations forms hairpin loops which participate in stabilizing protein binding.
    d. Most helix-turn-helix proteins bind with contacts on both sides of the DNA helix. Thus, inverted repeat DNA sequences are required to facilitate such contacts.
    answer: b

29. One example in which RNA acts as a transcription repressor is
    a. the gene that encodes CAP.
    b. the lac repressor gene.
    c. the lac Z, Y and A genes.
    d. the cI gene of bacteriophage λ.
    answer: a

Chapter 36   Regulation of Gene Expression in Eukaryotes

1. The specific attenuation mechanism that operates to regulate tryptophan biosynthesis in *E. coli* would not work in eukaryotes because
    a. mRNA turnover is much slower in eukaryotes.
    b. additional transcription factors are required for eukaryotic gene expression.
    c. in eukaryotes, transcription takes place in the nucleus, whereas translation occurs in the cytoplasm.
    d. eukaryotic genes do not utilize transcription terminators.
    answer: c

2. Eukaryotic gene transcription is associated with
    a. constitutive heterochromatin.
    b. facultative heterochromatin.
    c. euchromatin.
    d. Barr bodies.
    answer: c

3. The microscopic structure that corresponds to an inactive, highly-condensed X chromosome is called
    a. euchromatin.
    b. a Barr body.
    c. a polytene chromosome.
    d. a Philadelphia chromosome.
    answer: b

4. Which of the following individuals would have two Barr bodies in each of their cells?
    a. a Turner female (X).
    b. a normal female (XX).
    c. an XXY male.
    d. an XXX female.
    answer: d

5. In polytene chromosomes, the puffs are thought to be
    a. heterochromatic.
    b. correlated with actively transcribed genes.
    c. insensitive to hormones.
    d. highly methylated.
    answer: b

222

6. Some investigators believe that gene expression may be associated with redistribution of nucleosomes along the DNA duplex. One direct line of evidence that may support this hypothesis is:
   a. In some cases, the pattern of micrococcal nuclease digestion is different in active chromatin compared to inactive chromatin.
   b. Inactive chromatin is more highly condensed than active chromatin.
   c. Inactive chromatin has a higher level of CpG methylation than active chromatin.
   d. Genomic imprinting (which illustrates the possibility of parental-specific inheritance of particular traits) supports the hypothesis of nucleosome redistribution in highly active genes.
   answer: a

7. Of the following, the DNA binding motif utilized by homeodomain proteins most resembles
   a. leucine zippers.
   b. helix-turn-helix motifs.
   c. zinc fingers.
   d. histones.
   answer: b

8. Genomic imprinting leads to differential expression of a particular allele depending on whether that allele was inherited from the mother or the father. The molecular mechanism that leads to genomic imprinting may be
   a. the action of maternal effect genes.
   b. the action of homeotic genes.
   c. the action of transposable elements.
   d. the effects of parental-specific inheritance of DNA methylation patterns.
   answer: d

9. In yeast, there is a locus called HMRE that is adjacent to the storage location for the a mating type genes. If HMRE were deleted,
   a. the α mating type would be expressed.
   b. the a mating type would be expressed.
   c. diploid cells would be stimulated to undergo meiosis and sporulation.
   d. the haploid-specific gene, RME1 would be expressed.
   answer: b

10. Which of the following DNA locations contain transposable elements?
   a. HML$_\alpha$ and HMR$_a$
   b. SIR1-SIR4
   c. HMLE and HMRE
   d. RME1
   answer: a

11. In yeast, the synthesis of RME (a negative regulator of meiosis) is inhibited by
    a. a1.
    b. α2.
    c. a complex between α1 and α2.
    d. a complex between α2 and a1.
    answer: d

12. Which of the following statements about the differences between eukaryotic enhancers and prokaryotic transcription regulatory sequences is NOT TRUE?
    a. Enhancer sites may be several kilobases or more away form the regions they control, whereas prokaryotic regulatory sequences are located immediately upstream of the genes they regulate.
    b. Enhancers may be located downstream as well as upstream of the genes they regulate, whereas prokaryotic regulatory sequences are located immediately upstream of the genes they regulate.
    c. Enhansers may be effective in either orientation in the DNA molecule whereas whereas prokaryotic regulatory sequences are only effective in one orientation.
    d. Enhancers may be trans-acting elements, whereas prokaryotic regulatory sequences are always cis with respect to the genes that they regulate.
    answer: d

13. A number of different proposals have been advanced to explain how enhancers can influence transcription over long distances. Which of the following statements best describes the most strongly favored hypothesis?
    a. Proteins bound at enhancers may bind to proteins bound to promoter regions, causing the DNA to fold or loop to bring the enhancer-protein complex into close contact with the gene that it regulates.
    b. Transcription may be stimulated by the enhancer element functioning as an attachment point to structural components of the nucleus.
    c. The enhancer may be just the initial binding site for a factor required for transcription which then moves along the duplex to the transcription initiation point.
    d. Enhancers act by a combination of all of the mechanisms described above.
    answer: a

14. Steroid receptor proteins are a class of closely-related DNA binding proteins that each contain tandem DNA binding motifs called
    a. zinc fingers.
    b. homeodomains.
    c. helix-turn-helix motifs.
    d. leucine zippers.
    answer: a

15. The zinc ion in zinc finger proteins is bound by
    a. four amino acid residues that are either cys or his.
    b. glutamate or aspartate residues.
    c. leucine residues.
    d. DNA.
    answer: a

16. In yeast, the DNA binding site for the GAL4 protein is a
    a. HRE (hormone response element).
    b. promoter.
    c. UAS (upstream activating sequence).
    d. CCAAT box.
    answer: c

17. The yeast GAL4 protein has two functional domains: an activation domain and a DNA binding domain. This was demonstrated through the production of a lexA-GAL4 chimeric protein in which the DNA binding domain came from the lexA protein. Which of the following statements best describes the properties of this chimeric protein?
    a. The chimeric protein displayed the same transcription activation activity as the normal protein.
    b. The chimeric protein activated transcription only if the DNA binding site was changed to reflect the lexA recognition site.
    c. The chimeric protein activated transcription only if the DNA binding site was placed adjacent to the transcription initiation site.
    d. The chimeric protein activated transcription only if the protein was treated with a protease to remove the lexA domain.
    answer: b

18. Different transcripts from a single tropomyosin gene lead to a set of different polypeptide products through a process called
    a. attenuation.
    b. gene rearrangement.
    c. alternative splicing.
    d. antitermination.
    answer: c

19. The activation domain of the yeast GAL4 protein is an example of a general class of transcription activation domains that are
    a. acidic.
    b. glutamine rich.
    c. proline rich.
    d. leucine rich.
    answer: a

20. Eukaryotic cells utilize all of the following mechanisms for control of translation EXCEPT
    a. phosphorylation of eIF-1 and EF-2.
    b. inactivation of translation by double-stranded RNA.
    c. activation of translation by synthesis of ppGpp.
    d. differential recognition of open reading frames.
    answer: c

21. In rabbit reticulocytes, the amount of globin proteins is coordinated with the amount of available heme, in part, by a translation control mechanism that involves
    a. a double-stranded RNA activated inhibitor.
    b. heme inactivation of a kinase that would otherwise inactivate eIF-2.
    c. an attenuation mechanism the is sensitive to the amount of available heme.
    d. positive control by heme binding to an activator of eIF-2.
    answer: b

22. The earliest regulatory proteins in the Drosophila oocyte are supplied by
    a. homeotic genes.
    b. pair-rule genes.
    c. Gap genes.
    d. maternal effect genes.
    answer: d

23. The first step that leads to differentiation of anterior and posterior regions of the Drysophila embryo is a result of
    a. a gradient of concentrations of regulatory proteins, supplied by maternal cells, within the oocyte.
    b. differential effects of gravity on the material within the egg cytoplasm.
    c. differential organization of material within the oocyte initiated by pair-rule genes.
    d. alternative splicing of maternally-supplied transcripts.
    answer: a

24. The yeast SIR genes
    a. function as cis-acting silencers of sequences at the HMRa and HMRα locations.
    b. code for trans-acting regulatory factors that silence expression of the sequences at the HMRa and HMRα locations.
    c. contain transposable elements that move adjacent to the HMRa or HMRα loci to SILENCE expression of the yeast mating type sequences.
    d. contain transposable elements that move adjacent to the HMRa or HMRα loci to ENHANCE expression of the yeast mating type sequences.
    answer: b

25. Regulation of the yeast GNC4 gene (which in turn regulates genes for amino acid biosynthesis), is an example of
    a. translational control of transcription.
    b. alternative splicing.
    c. a homeodomain gene.
    d. regulation of translation through covalent modification of translation proteins.
    answer: a

26. True or False? The arrangement of genes within the ANC-C locus of the Dysophila genome is co-linear with the body plan of the fly.

    answer: true

27. True or False? Proteins that bind to DNA in prokaryotes are almost always symmetrical, whereas eukaryotic transcriptional regulators are often asymmetrical.

    answer: true

28. Gal80 is a yeast repressor protein that prevents transcription of the gal1 gene by
    a. binding to DNA in the presence of galactose.
    b. binding to DNA in the absence of galactose.
    c. binding to the gal4 protein in the presence of galactose.
    d. binding to the gal4 protein in the absence of galactose.
    answer: d

29. In Drosophila, the eve gene codes for a protein that is expressed in seven stripes along the length of the developing embryo, but not in the locations between the stripes. Control of this pattern of alternating expression and non-expression is achieved by
    a. a gradient in the concentration of a single critical regulatory protein that forms along the length of the embryo.
    b. negative regulation by pair-rule genes.
    c. differential expression of eve from transcripts that are spliced differently in the different stripes.
    d. translational control of gene expression that depends on different eIF-2 phosphorylation levels in the different stripes.
    answer: b

30. Eukaryotic transcription regulators often work in tandem. One example of a binding mechanism that is used to hold two such partner proteins together is
    a. leucine zippers.
    b. zinc fingers.
    c. helix-turn-helix domains.
    d. covalent bonding.
    answer: a

31. If a teratoma cell is implanted into an early embryo,
    a. a teratoma develops within the embryo.
    b. the embryo fails to differentiate.
    c. only differentiated cells revert back to normal.
    d. stem cells can revert back to normal.
    answer: d

Chapter 37 Immunobiology

1. The two classes of lymphocytes are
   a. T cells and macrophages.
   b. helper T cells and killer T cells.
   c. B cells and T cells.
   d. macrophages and monocytes.
   answer: c

2. The cell-mediated immune response
   requires only T cells.
   a. requires B cells and macrophages.
   b. requires B cells and T cells.
   c. requires antibody-secreting B cells.
   answer: a

3. In IgG antibodies
   a. the heavy chains are constant whereas the light chains convey variability for all of the different specific antibodies.
   b. the light chains are constant whereas the heavy chains convey variability for all of the different specific antibodies.
   c. both the heavy chains and light chains each have constant sequences at their termini ends and variable sequences at their COOH termini.
   d. both the heavy chains and light chains each have constant sequences at their COOH termini and variable sequences at their amino termini.
   answer: c

4. IgG and IgM are different in that
   a. IgG is an aggregate of 5 tetramers, whereas IgM is a single tetramer.
   b. IgG is secreted whereas some forms of IgM remain within the B cell membrane.
   c. For a particular antigenic specificity, only the light chains are different between the two classes of immunoglobulins.
   d. IgG is synthesized first, followed several weeks later by IgM.
   answer: b

5. For different IgG molecules, antibody diversity is generated by
   a. alternative splicing of the RNA transcripts.
   b. highly regulated expression of the 1000's of different antibody genes.
   c. post translational modification of antibody genes.
   d. somatic recombination and somatic mutation of antibody genes.
   answer: d

6. True or False?   There are two gene clusters that code for antibody light chains, whereas only one gene cluster codes for all of the different types of heavy chains.

    answer: true

7. The genes for IgG differ from their corresponding IgM genes only in
    a. the light chains.
    b. the V region of the heavy chains.
    c. the J region of the heavy chains.
    d. the C region of the heavy chains.
    answer: d

8. Of the following events leading to antibody production, which occurs first?
    a. the B cell attempts to produce functional λ light chain.
    b. the B cell attempts to produce functional κ light chain.
    c. V_H-D-J_H joining occurs.
    d. there is a class switch from membrane-bound to secreted antibody production.
    answer: c

9. The number of different heavy chain C genes is
    a. 1
    b. 4
    c. 8
    d. >1000
    answer: c

10. Of the following, which can be the result of alternative RNA splicing?
    a. the difference between κ and λ light chains.
    b. the switch between membrane-bound and the secreted form of IgM.
    c. the generation of somatic mutations in the V_H region.
    d. the generation of somatic mutations in the framework regions of the antibody genes.
    answer: b

11. The structure of light chain genes is
    a. V-J-D-C
    b. C-J-D-V
    c. V-D-J-C
    d. V-J-C
    answer: d

12. B cells are triggered to divide
    a. when they switch from IgM to IgG production.
    b. by CAP formation.
    c. when they respond to γ-interferon secreted by macrophages.
    d. when they become memory cells.
    answer: b

13. When stimulated by T cells, macrophages can produce cytotoxic chemicals that include
    a. γ-interferon.
    b. interleukins I and II.
    c. tumor necrosis factor and nitric oxide.
    d. β microglobin
    answer: c

14. A rare inherited disease called DiGeorge syndrome is associated with an underdeveloped thymus. Thus, individuals affected by this disease suffer from
    a. an inability to produce macrophages.
    b. an impaired cellular immune response.
    c. an impaired humoral immune response.
    d. both an impaired cellular immune response and an impaired humoral immune response.
    answer: b

15. The principal determinants of self-nonself recognition are
    a. the major histocompatability complex.
    b. the complement system.
    c. B cells.
    d. macrophages.
    answer: a

16. Cytotoxic T cells
    a. lyse bacterial cells.
    b. lyse host cells that have foreign antigens attached to their surface MHC proteins.
    c. stimulate macrophages to phagocytize foreign cells.
    d. stimulate B cells to produce antibodies.
    answer: b

17. The complement system is activated by
    a. cytotoxic T cells.
    b. helper T cells.
    c. macrophages.
    d. IgG-antigen complexes.
    answer: d

18. Which of the following statements most accurately describes T cell recognition of antigens?
    a. T cells recognize antigens only when the antigen is associated with an MHC protein on the surface of another host cell.
    b. T cells recognize antigens only when the antigen is present on the surface of a foreign cell.
    c. T cells recognize antigens by essentially the same process as antigen recognition by B cells.
    d. T cells recognize antigens only when stimulated to do so by B cells.
    answer: a

19. The molecular mass of an IgG molecule is about
    a. 60,000 daltons.
    b. 150,000 daltons.
    c. 900,000 daltons.
    d. 3,000,000 daltons.
    answer: b

20. κ and λ are
    a. V_H genes.
    b. V_C genes.
    c. MHC genes.
    d. antibody light chain genes.
    answer: d

21. In comparing a primary immune response to a secondary immune response, you would note that
    a. the response time is approximately the same for the appearance of both IgM and IgG. However, the maximum serum level of IgG is higher for the secondary response then for the primary response.
    b. there is a faster appearance of both IgM and IgG and higher maximum serum levels for a secondary immune response than for a primary immune response.
    c. there is a faster appearance of both IgM and IgG for a secondary immune response than for a primary immune response. However, the maximum serum levels of IgM are approximately the same for the primary and secondary immune response.
    d. there is a faster appearance of IgG for a secondary immune response than for a primary immune response. However, IgM is not produced during a secondary immune response.
    answer: c

232

22. Both B cells and T cells are produced in the
    a. thymus.
    b. spleen.
    c. lymph nodes.
    d. bone marrow.
    answer: d

23. Which of the following is NOT one of the steps needed to activate specific B cells to proliferate?
    a. an antigne binds to IgM on the surface of a naïve B cell.
    b. the antibody-antigen complex is internalized in the cell and degraded by proteolysis.
    c. the B cell binds to a macrophage that recognizes MHC-II proteins bound to antigen fragments on the surface of the B cell.
    d. secreted interleukin-2 and other polypeptides stimulate the B cell to proliferate.
    answer: c

24. Which of the following is NOT considered an antigen-presenting cell?
    a. macrophages
    b. B cells
    c. dendritic cells
    d. cytotoxic T cells.
    answer: d

25. The time delay between antigen stimulation of a specific B cell and the appearance of IgG antibodies in the blood serum is about
    a. 10-20 minutes.
    b. 24 hours.
    c. 4-7 days.
    d. 28 days.
    answer: c

26. Vaccination works to confer immunity by stimulating the production of
    a. cytotoxic T cells.
    b. supressor T cells.
    c. macrophages.
    d. memory cells.
    answer: d

27. T cell diversity is generated by a process that involves
    a. alternative RNA splicing.
    b. gene rearrangements similar to those involved in antibody production.
    c. somatic mutation rather than the gene rearrangements involved in antibody production.
    d. a combination of alternative splicing, gene rearrangements and somatic mutations.
    answer: b

28. Class I MHC proteins are found
    a. on almost all cell membranes.
    b. only on T cell membranes.
    c. primarily on antigen-presenting cell membranes.
    d. only on B cell membranes.
    answer: a

29. Class II MHC proteins are found
    a. on almost all cell membranes.
    b. only on T cell membranes.
    c. primarily on antigen-presenting cell membranes.
    d. only on B cell membranes.
    answer: c

30. The CD family of proteins
    a. are T cell receptor proteins.
    b. are members of the MHC complex family of proteins.
    c. are cell-cell adhesion proteins.
    d. are part of the complement system.
    answer: c

Chapter 38 Cancer and Carcinogenesis

1. Most human cancers are caused by
   a. hereditary factors
   b. environmental factors
   c. viruses
   d. bacteria
   answer: b

2. Burkitt's lymphoma is associated with
   a. the Hepatitis B virus
   b. the Epstein-Barr virus
   c. chromosome translocations
   d. both the Epstein-Barr virus and chromosome translocations
   answer: d

3. Chromosome translocations can lead to cancer through
   a. activation of tumor supressor genes.
   b. mutation of protooncogenes.
   c. juxtaposition of a protooncogene next to a gene that is normally transcribed.
   d. hypersensitization of the affected cell to harmful environmental factors.
   answer: c

4. Retroviruses function in carcinogenesis by
   a. making a DNA copy of their genetic material that is integrated into the host genome.
   b. initiating a lytic cycle.
   c. inhibiting DNA synthesis in the host cell.
   d. inhibition of the host genome using antisense RNA.
   answer: a

5. Tumor-associated mutant Ras proteins
   a. are expressed at a much higher level in tumor cells than in normal cells.
   b. do not bind GTP.
   c. bind GTP but do not have GTP hydrolysis activity.
   d. do not associate with the plasma membrane.
   answer: c

6. Which of the following statements about retroviral oncogenes is correct?
   a. They usually do not contain introns.
   b. They are usually not transcribed.
   c. They are recessive with respect to their normal cell counterpart..
   d. They have very little sequence homology with their normal cell protooncogene counterparts.
   answer: a

7. The major cause of human liver cancer is
   a. smoking.
   b. prior infection with the hepatitis B virus.
   c. a high-fat diet.
   d. occupational exposure to chemical carcinogens.
   answer: b

8. Which of the following is a true statement?
   a. Oncogenes are usually dominant with respect to their corresponding protooncogenes, whereas mutant tumor supressor genes are usually recessive to their normal counterparts.
   b. Oncogenes are usually recessive with respect to their corresponding protooncogenes, whereas mutant tumor supressor genes are usually dominant to their normal counterparts.
   c. Both oncogenes and tumor supressor mutations are usually dominant with respect to their normal counterparts.
   d. Both oncogenes and tumor supressor mutations are usually recessive with respect to their normal counterparts.
   answer: a

9. The retinoblastoma gene is an example of a
   a. protooncogene.
   b. cellular oncogene.
   c. viral oncogene.
   d. tumor supressor gene mutation.
   answer: d

10. The most common gene mutation associated with human cancers affects
    a. ras
    b. c myc
    c. p53
    d. src
    answer: c

11. c myc utilizes all of the following DNA-binding motifs except a
    a. leucine zipper.
    b. helix-turn-helix.
    c. highly basic region.
    d. zinc finger.
    answer: d

12. c myc interacts with a partner protein called
    a. max
    b. fos
    c. jun
    d. AP-1
    answer: a

13. The normal function of most protooncogenes is
    a. supression of the rate of cell division.
    b. initiation of programmed cell death (apoptsis).
    c. cell cycle regulation
    d. defense against viral infection
    answer: c

14. Which of the following statements about p53 is NOT CORRECT?
    a. p53 arrests the cell cycle at the G1/S border.
    b. p53 can induce programmed cell death (apoptsis).
    c. p53 binds to DNA and functions as a transcription activator.
    d. normal p53 is essential for embryonic development.
    answer: d

15. Which of these statements is NOT CORRECT?
    a. Cancer cells usually contain a mutation in a cell cycle gene.
    b. Cancer usually results from a single mutational event.
    c. Most tumors are clonal descendants of a single cell.
    d. Most cancers are not hereditary.
    answer: b

16. Protooncogenes can become oncogenes in all of the following ways EXCEPT,
    a. mutation.
    b. chromosome translocation.
    c. formation of a fusion protein.
    d. deletion.
    answer: d

17. The inherited gene in bilateral retinoblastoma is recessive with respect to its normal counterpart. Usually, recessive traits are not phenotypically expressed unless the individual is homozygous for the recessive gene. However, in this case, up to 90% of the children who inherit the mutant RB allele develop tumors in both eyes even though they are heterozygous. One possible explanation for this apparent paradox is
    a. Tumors don't develop until a second somatic mutation occurs.
    b. Expression of the mutant RB protein inactivates the normal RB protein.
    c. The normal RB gene is not expressed.
    d. The gene is sex-linked.
    answer: a

18. For an oncogene that corresponds to a polypeptide growth factor, you would expect that
    a. overexpression of the normal growth factor would reverse the effects of the oncogene.
    b. only cells that express the growth factor receptor protein would be susceptible to the oncogene.
    c. the oncogene is probably a G protein.
    d. the oncogene probably has tyrosine kinase activity.
    answer: b

19. Oncogenes that have tyrosine kinase activity
    a. probably correspond to protooncogenes that are growth factor receptors.
    b. probably correspond to protooncogenes that are polypeptide growth factors.
    c. are probably G proteins.
    d. are probably DNA binding proteins.
    answer: a

20. Von Hippel-Lindau disease is an inherited condition that predisposes affected individuals to develop tumors in one or more of many different tissues. If the mutation that causes this disease is a deletion, you would expect the corresponding normal gene to be
    a. an oncogene.
    b. a protooncogene.
    c. a tumor supressor gene.
    d. a DNA repair gene.
    answer: c

238

21. Recently, a gene called BRCA1 was identified as one of the links to the inherited form of breast cancer. In addition, two somatic mutations in BRCA1 can lead to breast tumors. The normal form of this gene most likely codes for
    a. an oncogene.
    b. a protooncogene.
    c. a tumor supressor gene.
    d. a DNA repair gene.
    answer: c

22. The normal activity of protooncogenes is to
    a. inhibit integration of viral DNA into the host genome.
    b. mediate programmed cell death.
    c. prevent cells from multiplying in mature, fully-differentiated tissues.
    d. promote regulated cell growth and division.
    answer: d

23. Human cervical cancers are common in women who have been infected with
    a. hepatitis B virus.
    b. papilloma viruses.
    c. Epstein-Barr virus.
    d. the Rous sarcoma virus.
    answer: b

24. Which of these statements about tumor cells in culture is NOT CORRECT?
    a. Cultured tumor cells divide indefinitely.
    b. Cultured tumor cells lack contact inhibition.
    c. Cultured tumor cells require a solid surface or anchor on which to grow.
    d. Cultured tumor cells will often induce tumors when injected into an individual of their parent species.
    answer: c

25. Chronic myelogenous leukemia is associated with a chromosomal abnormality called
    a. Trisomy 21.
    b. Burkitt's lymphoma.
    c. the Philadelphia chromosome.
    d. cmyc.
    answer: c

26. A mutation of a particular growth factor receptor protein results in tyrosine kinase activity even in the absence of the appropriate growth factor. If this mutation led to tumor formation, it would be called
    a. a tumor supressor gene.
    b. an anti-oncogene.
    c. a protooncogene.
    d. an oncogene.
    answer: d

27. The p53 gene product is
    a. a G protein.
    b. a DNA-binding protein.
    c. a tyrosine kinase.
    d. a serine/threonine kinase.
    answer: b

28. One of the enzymes required by RNA tumor viruses to integrate their genetic material into the host genome is
    a. DNA polymerase.
    b. GTPase.
    c. reverse transcriptase.
    d. protease.
    answer: c

29. The tumor causing genes in DNA viruses
    a. are oncogenes that are related to protooncogenes that are normally present in the host cell.
    b. are viral-specific genes that promote expression of early viral genes, often through interaction with host tumor supressor proteins.
    c. are tumor supressor genes.
    d. are essentially identical in structure to genes in RNA viruses.
    answer: b

Chapter 39 The Human Immunodeficiency Virus (HIV) and Acquired Immunodeficiency Syndrome (AIDS)

1. The number of people worldwide that are currently infected by HIV is
   a. about 1.5 million.
   b. about 5 million.
   c. about 10 million.
   d. over 2% of the world's population.
   answer: d

2. In Sub-Saharan Africa,
   a. HIV infection is rare in women.
   b. About four times more men than women are infected with HIV.
   c. HIV infects about equal numbers of men and women.
   d. the incidence of HIV infection has risen so fast in women that now, the number of women infected with HIV is about 1.5 time greater than the number of infected men.
   answer: c

3. Some investigators have questioned whether HIV infection actually causes AIDS. This belief is supported by the only one of the following statements that is TRUE.
   a. Less than 20% of those who have been infected by HIV for more than 10 years have progressed to AIDS.
   b. Investigations in animal systems have not uncovered any closely-related viruses that infect other animals.
   c. There have been cases of severe acquired immune deficiency who have tested negative for both HIV-1 and HIV-2.
   d. Drugs that inhibit HIV proliferation have no effect on AIDS.
   answer: c

4. HIV-2 exhibits the greatest sequence homology with
   a. HIV-1.
   b. HTLV-1.
   c. HTLV-2.
   d. SIV.
   answer: d

5. A typical time-span between the appearance of anti HIV antibodies in a patient's blood stream and evidence of AIDS symptoms is
   a. 4-8 weeks.
   b. 2-3 years.
   c. up to 12 years.
   d. 12-20 years.
   answer: c

6. HIV infects
   a. all human blood cells.
   b. all human cells.
   c. primarily CD4$^+$ helper T cells and macrophages.
   d. only CD4$^+$ helper T cells.
   answer: c

7. High levels of circulating p24 antigen is a sign of
   a. initial infection. (It can be detected within a few weeks of infection.)
   b. viral latency.
   c. a favorable response to antiviral drugs.
   d. the onset of the terminal phase of the disease.
   answer: d

8. HIV can induce cell death of infected CD4$^+$ cells through at least two mechanisms. One of these is
   a. inducing cell-to-cell fusion in which one infected CD4$^+$ cell can fuse with up to several hundred uninfected CD4$^+$ cells.
   b. shut down of host cell protein production caused by integration of the HIV genome into the host cell genome.
   c. viral-directed production of reverse transcriptase which is toxic to the host cells.
   d. viral-directed production of an oncogene product.
   answer: a

9. The promoter for transcription of RNA copies of the HIV genome is located
   a. upstream of the integration site in the host genome.
   b. in the LTR region of the viral genome.
   c. in the VPR region of the viral genome.
   d. in the tat region of the viral genome.
   answer: b

10. Viral entry into the host cell involves interaction between the CD4 receptor and a viral protein called
    a. the HIV protease.
    b. integrase.
    c. gp120
    d. gp160.
    answer: c

11. The viral coded NF-kB enhanser activates several T cell genes as well as viral genes. These include
   a. IL-2 and the high-affinity IL-2 receptor.
   b. the CD4 receptor.
   c. Sp1.
   d. RNA polymerase.
   answer: a

12. The late Jonas Salk and his coworkers proposed that a low dose rather than a high dose vaccine be developed for prevention of HIV infection. Their reasoning was as follows:
   a. The HIV virus may be pathogenic because it overloads the immune system. Thus, a low dose that does not overload the immune system would be preferable to a high dose vaccine.
   b. A low dose would target stimulation of the cellular immune response which is more effective against HIV than the humoral immune response, whereas a high dose vaccine would elicit an antibody response at the expense of the cellular response.
   c. A low dose would target stimulation of the humoral immune response which is more effective against HIV than the cellular immune response, whereas a high dose vaccine would not elicit an antibody response.
   d. A low dose vaccine would target both Th1 and Th2 cells.
   answer: b

13. Three genes are characteristic of most retroviruses. These are:
   a. gag, pol, env.
   b. gp120, p24, p9.
   c. ltr, pol, rev.
   d. gag, rev, nef.
   answer: a

14. True or False? The HIV virus genome contains several overlapping genes that are translated in different reading frames.

   answer: true

15. The pol polyprotein is the precursor to all of the following viral proteins EXCEPT:
   a. gp160.
   b. reverse transcriptase.
   c. protease.
   d. integrase.
   answer: a

16. Over half of those individuals diagnosed with AIDS contract Karposi's sarcoma. This is evidence that
    a. the HIV virus carries an oncogene.
    b. the immune system normally protects humans against cancer.
    c. HIV causes Karposi's sarcoma.
    d. individuals with early stages of Karopsi's sarcoma are particularly susceptible to AIDS.
    answer: b

17. Recently, several drug companies have developed new drugs for AIDS that inhibit the HIV protease. One of the effects of inhibiting the viral protease would be to
    a. prevent RNA transcription.
    b. prevent processing of the gag polyprotein to produce p24, p7, p9 and p17.
    c. inhibit reverse transcriptase.
    d. prevent integration of the viral genome into the host DNA.
    answer: b

18. In a cell infected by HIV, which of the following proteins is coded by the host cell genome?
    a. gp120.
    b. reverse transcriptase.
    c. integrase.
    d. Ets-1.
    answer: d

19. Complex carbohydrates are added to the viral glycoprotein
    a. by the action of the viral protein, tat.
    b. while the glycoprotein is on the surface of the host cell membrane.
    c. in the host cell Golgi apparatus.
    d. after the virus buds from the surface of the host cell.
    answer: c

20. HIV was first identified as the causal agent for AIDS by
    a. Jonas Salk.
    b. Tim Mosmann and Robert Coffman.
    c. M. A. Nowak and A. J. McMichael.
    d. Luc Montagnier and Robert Gallo.
    answer: d

21. It could be argued that HIV-2 should be easier to study than HIV-1 because
    a. HIV-2 is much more common in the United States than HIV-1.
    b. Researchers would not need to take safety containment precautions when handling HIV-2.
    c. HIV-2 has a broader host range than HIV-1, thus facilitating the development of animal models in which to study the disease.
    d. HIV-2 has a high sequence homology to HTLV-2 which is one of the best-studied human viruses.
    answer: c

22. Which of the following statements about retroviruses is FALSE?
    a. They all contain RNA as their genetic material.
    b. They all use reverse transcriptase to make double-stranded DNA copies of their single-stranded RNA genome.
    c. They all carry the same 10 minimum essential genes.
    d. They all function by integrating their genetic material into the host genome.
    answer: c

23. True or False? HIV is considered a transforming virus.

    answer: false

24. True or False? The human immune system is unable to produce antibodies against HIV proteins.

    answer: false

25. During the progression from initial infection to the diagnosis of AIDS to eventual death, the number of cytotoxic T lymphocytes
    a. decreases steadily.
    b. remains high even during latency periods.
    c. parallels the patient's viral titer.
    d. remains low until the development of full-blown AIDS.
    answer: b

26. At the time of seroconversion,
    a. the patient's viral titer is greatly increased.
    b. there is an increase in the appearance of p24 in the serum.
    c. IgM antibodies to HIV proteins can be detected.
    d. there is a general collapse of the patient's immune system.
    answer: c

27. Which of the following statements about gp160 is TRUE?
    a. gp160 is produced through the fusion of gp120 and gp41.
    b. Gp160 is inactive until it is cleaved to form gp120 and gp41.
    c. gp160 is a product of the gag gene.
    d. gp160 is one of the nucleocapsid core proteins.
    answer: b